Builder's Estimating Databook

Robert L. Taylor and S. Blackwell Duncan

TAB BOOKS

Blue Ridge Summit, PA

FIRST EDITION
FIRST PRINTING

Copyright © 1990 by TAB BOOKS
Printed in the United States of America

Library of Congress Cataloging in Publication Data

Taylor, Robert L. (Robert Lowell), 1903-
    Builder's estimating databook / by Robert L. Taylor and S.
Blackwell Duncan.

    p.   cm.
    ISBN 0-8306-1368-4   ISBN 0-8306-2768-5 (pbk.)
    1. Building—Estimates.   I. Duncan, S. Blackwell.   II. Title.

TH435.T24   1989   89-36596
692'.5—dc20   CIP

TAB BOOKS offers software for sale. For information and a catalog, please contact TAB Software Department, Blue Ridge Summit, PA 17294-0850.

Questions regarding the content of this book
should be addressed to:

    Reader Inquiry Branch
    TAB BOOKS
    Blue Ridge Summit, PA 17294-0214

Book Editor: Cherie R. Blazer
Production: Katherine Brown
Cover Design: Lori E. Schlosser
Cover Photograph: Reprinted by permission of The Ryland Group, Inc., Builders of Ryland Homes.

Contents

M   *152*

T   *283*

Appendix   **335**

Glossary   **254**
Index   **272**

Introduction

While it is always necessary to do some actual scaling or measuring of plans or projects to do accurate estimating, a great many shortcuts will be found in *Builder's Estimating Databook*. Quantities may be quickly determined using the many tables provided. For example, the areas of rooms of various sizes and their cubic contents are included, as well as quick charts for brick, concrete, masonry, and many other items.

This is the most complete, concise, and easy-to-use book ever compiled and published for general use in the building industry. It is prepared in a very simplified form in order to prove helpful for anyone who is interested in any phase of building.

The proper and detailed use of the complete easy-to-use index is an indication of the thoroughness and simplicity of the hundreds of tables in the book. The index covers a wide range of building products and materials. Building terms, parts, and component parts are indicated by the names generally used in the industry. As a result, it is often necessary to refer to various sections of the index and book. For example, Roofing may be found under this name or under type of roofing, such as: Asphalt, Wood Shingles, etc.

This book has been prepared and written for general use. It will benefit the individual who is interested in building his own home, the present home owner, investor, builder, architect, designer, contractor, as well as the craftsman, student, dealer, salesman, manufacturer and producer. Anyone interested in any phase of the building industry will find it useful and helpful in a great many respects.

Starting with excavation, foundations, floors, framing, roofs, chimneys, etc., complete and detailed information for estimating, comparing materials, physical properties, sizes, weights, packing, and many other very important and useful tables and charts may be found that answer many questions on materials and building products.

Along the lines of design, this book contains information on general sizes, types, specifications, weights, load limits, strengths, symbols, abbreviations, comparative values, etc.

AGGREGATE SIZES—TOTAL PERCENT PASSING
OF SQUARE SCREEN OPENINGS

	4″	3½″	3″	2½″	2″	1½″	1″	¾″	⅜″
No. 1	100	90 – 100	35 – 70	0 – 15					
No. 12	100	90 – 100	65 – 85	25 – 60		0 – 15			
No. 2			100	90 – 100	35 – 70	0 – 15			
No. 3					100	80 – 100	20 – 60	0 – 20	0 – 5

	2″	1½″	1″	¾″	½″	⅜″	¼″	⅛″
No. 34	100	90 – 100		30 – 65		0 – 20		
No. 4			100	80 – 100	20 – 60	5 – 30		
No. 46			100	95 – 100	65 – 90	35 – 65		
No. 6					100	90 – 100		
No. 7							100	
No. 6a				100	90 – 100	40 – 70		
No. 6b					100	75 – 100		
No. 9						100	95 – 100	
No. 9c						100	90 – 100	

AIR INFILTRATION BARRIER
(HOUSEWRAP)—PROPERTIES

Packaged in rolls:

- Widths (ft.)—3, 4½, 9
- Lengths (ft.)—111, 165, 195, 222
- Bursting strength—100 – 125 psi
- Weight (lbs.)—8.5 – 16.5 (1000 sq. ft.)
- Thickness—6 – 10 mils

ALUMINUM—MISCELLANEOUS STOCK ITEMS (SIZES)

Items	Sizes (In.)
Decorative Panels—Perforated Types, Various Patterns	24 × 36 and 36 × 36, 24 gauge
Rods	$3/8$ (dia.), 72 (length)
Angles	1 × 1 × $1/16$ $3/4$ × $3/4$ × $1/8$ 72 and 96 (lengths)
Tubing (OD) Dimensions	$3/4$, 1, 1 $1/4$ 72 and 96 (lengths)
Bars	$3/4$ × 1 $1/8$ × 72 1 × $1/4$ × 72

ANCHORS—SCREW

Lead				Fiber		
Anchor Length (In.)	Screw Screw Size	Dia. (In.)	Drill Size	Screw Size	Anchor Length (In.)	Screw Length (In.)
$3/4$	6 – 8	$1/8$ – $3/16$	$1/4$	5 – 6	$5/8$ – 1	1 – 1 $1/2$
1 $1/2$	6 – 8	$1/8$ – $3/16$	$1/4$	7 – 8	$5/8$ – 1 $1/2$	1 – 2 $1/2$
1	10 – 14	$3/16$ – $1/4$	$5/16$	9 – 10	$3/4$ – 1 $1/2$	1 $1/4$ – 2 $1/2$
1 $1/2$	10 – 14	$3/16$ – $1/4$	$5/16$	11 – 12	$3/4$ – 1 $1/2$	1 $1/4$ – 2 $1/2$
1	16 – 18	$5/16$	$3/8$	14	1 – 2	1 $1/2$ – 3 $1/2$
1 $1/2$	16 – 18	$5/16$	$3/8$	16	1 – 2	1 $1/2$ – 3 $1/2$
1 $3/4$	20 – 24	$3/8$	$7/16$	20	1 – 2	1 $1/2$ – 3 $1/2$

Numerous sizes and types available

ANCHORS—SIZES

Type	Sizes (In.)
Bolts:	$3/8 - 1/2 \times 6$ to 24 $5/8 \times 8$ to 24 $3/4 \times 8$ to 24
Door Buck:	$1/8 \times 1 \times 6$ $3/16 \times 1 1/2 \times 8$ Legs 1 & $1 1/2$
Joist: (Style T)	$1/8 \times 1 \times 15$ & 20 $3/16 \times 1 \times 15$ & 20 $1/8 \times 1 1/2 \times 15$ & 20
(Style L)	$1/8 \times 1 \times 15$ & 20 $3/16 \times 1 \times 15$ & 20 $3/16 \times 1 1/2 \times 15$ & 20
Stone:	$3/16 \times 1 \times 6$ $3/16 \times 1 \times 8$ $3/16 \times 1 \times 10$ Legs 1 & $1 1/2$

ANGLES—STEEL (RECOMMENDED SIZES, MASONRY OPENINGS

Opening Size (Ft.)	Angle Size (In.)
Up to 3 wide	$3 1/2 \times 3 1/2 \times 1/4$
3 – 7 wide	$3 1/2 \times 3 1/2 \times 5/16$
7 and 8 wide	$3 1/2 \times 3 1/2 \times 3/8$
8 – 10 wide	$5 \times 3 1/2 \times 3/8$

ANGLES—STEEL (SIZES, WEIGHTS)

Per Lineal Foot in Pounds

Sizes (In.)	1/8	3/16	1/4	5/16	3/8	7/16	1/2	9/16	5/8	11/16	3/4	13/16	7/8	15/16	1
3/4 × 3/4	.59	.84	....	....	....	....	....	....	....	....	....	....	....	....	....
1 × 1	.80	1.16	1.49	....	....	....	....	....	....	....	....	....	....	....	....
1 1/4 × 1 1/4	1.01	1.48	1.92	....	....	....	....	....	....	....	....	....	....	....	....
1 1/2 × 1 1/2	1.23	1.80	2.34	2.86	3.35	....	....	....	....	....	....	....	....	....	....
1 3/4 × 1 3/4	1.44	2.12	2.77	3.39	4.0	....	....	....	....	....	....	....	....	....	....
2 × 1 1/2	1.44	2.12	2.77	3.39	3.99	....	....	....	....	....	....	....	....	....	....
2 × 2	1.65	2.44	3.19	3.92	4.7	5.3	....	....	....	....	....	....	....	....	....
2 1/4 × 2 1/4	1.86	2.75	3.62	4.5	5.3	6.1	6.0	....	....	....	....	....	....	....	....
2 1/2 × 2	1.86	2.75	3.62	4.5	5.3	6.1	6.8	....	....	....	....	....	....	....	....
2 1/2 × 2 1/2	2.08	3.07	4.1	5.0	5.9	6.8	7.7	....	....	....	....	....	....	....	....
3 × 2	....	3.07	4.1	5.0	5.9	6.8	7.7	....	....	....	....	....	....	....	....
3 × 2 1/2	....	3.39	4.5	5.6	6.6	7.6	8.5	9.5	....	....	....	....	....	....	....
3 × 3	2.50	3.70	4.9	6.1	7.2	8.3	9.4	10.4	11.5	....	....	....	....	....	....
3 1/2 × 2 1/2	....	....	4.9	6.1	7.2	8.3	9.4	10.4	11.5	12.5	13.4	....	....	....	....
3 1/2 × 3	....	....	5.4	6.6	7.9	9.1	10.2	11.4	12.5	13.6	14.7	....	....	....	....
3 1/2 × 3 1/2	....	....	5.8	7.2	8.5	9.8	11.1	12.4	13.6	14.8	16.0	15.8	16.8	....	....
4 × 3	....	....	5.8	7.2	8.5	9.8	11.1	12.4	13.6	14.8	16.0	17.1	18.3	....	....
4 × 3 1/2	....	....	6.2	7.7	9.1	10.6	11.9	13.3	14.7	16.0	17.1	....	....	....	....
4 × 4	....	....	6.6	8.2	9.8	11.3	12.8	14.3	15.7	17.1	18.5	19.9	21.2	....	....
5 × 3	....	....	....	8.2	9.8	11.3	12.8	14.3	15.7	17.1	18.5	19.9	21.2	....	....
5 × 3 1/2	....	....	....	8.7	10.4	12.0	13.6	15.2	16.8	18.3	19.8	21.3	22.7	24.2	....
5 × 4	....	....	....	9.8	11.0	12.8	14.5	16.2	17.8	19.5	....	....	....	....	....
5 × 5	....	....	....	10.3	12.3	14.3	16.2	18.1	20.0	21.8	....	....	....	....	....
6 × 3 1/2	....	....	9.8	....	11.7	13.5	15.3	17.1	18.9	20.6	22.4	24.0	25.7	27.3	28.9
6 × 4	....	....	10.3	....	12.3	14.3	16.2	18.1	20.0	21.8	23.6	25.4	27.2	28.9	30.6
6 × 6	....	....	....	....	14.9	17.2	19.6	21.9	24.2	26.5	28.7	31.0	33.1	35.3	37.4
7 × 4	....	....	....	....	13.6	15.8	17.9	20.0	22.1	24.2	26.2	28.2	30.2	32.1	34.0
8 × 4	....	....	....	....	....	17.2	19.6	21.9	24.2	26.5	28.7	31.0	33.1	35.3	37.4
8 × 6	....	....	....	....	....	20.2	23.0	25.7	28.5	31.2	33.8	36.5	39.1	41.7	44.2
8 × 8	....	....	....	....	....	....	26.4	29.6	32.7	35.8	38.9	42.0	45.0	48.1	51.0

B

BEAMS—CANTILEVERED WOOD (MATERIALS REQUIRED)

Span (Ft.)	2 × 8 (In.) Length No.	(Ft.)	2 × 6 (In.) Length No.	(Ft.)	2 × 4 (In.) Length No.	(Ft.)	Total F.B.M.	2 × 6 (In.) Length No.	(Ft.)	2 × 4 (In.) Length No.	(Ft.)	Total F.B.M
20			4	12	2	8	59	4	12	2	10	62
22			4	14	2	10	70	4	14	2	10	70
24			4	14	2	10	70	4	14	2	12	72
26			4	16	2	10	78	4	16	2	12	80
28	2	16	2	16	2	12	91	4	16	2	12	80
30	2	18	2	18	2	12	100	4	18	2	14	91
32	2	20	2	18	2	12	100	2	20	2	14	95

BEAMS—LIGHTWEIGHT STEEL (SIZES, WEIGHTS)

Depth (In.)	Width	Web Thickness	Flange Thickness	Wt./ Lin. Ft. (Lbs.)
6	1 7/8	1/8	3/16	4.5
7	2 1/8	1/8	3/16	5.6
8	2 1/4	1/8	3/16	6.6
10	2 3/4	3/16	3/16	8.9
12	3	3/16	1/4	11.7

Many other sizes available

BEAMS—PRECAST CONCRETE

Width (In.)	Thickness (In.)	Max Lengths	Wt. (Lbs.)/ Sq. Ft.
12	6	22	40
12	6 5/16	22	45
15 15/16	6 5/16	22	50
16	8	26	55

Specific products will vary

BEAMS—STEEL I (SIZES, WEIGHTS)

Depth (In.)	Wt. Lbs. per Ft.	Flange Width	Web Thickness	Lengths (Ft.)
3	5.7	2.330	.170	5 – 40
	7.5	2.509	.349	5 – 40
4	7.7	2.660	.190	5 – 40
	9.5	2.796	.326	5 – 40
5	10.0	3.000	.210	5 – 60
	14.75	3.284	.494	5 – 60
6	12.5	3.330	.230	5 – 60
	17.25	3.565	.465	5 – 60
7	15.3	3.660	.250	5 – 60
	20.0	3.660	.450	5 – 60
8	18.4	4.000	.270	5 – 60
	23.0	4.171	.441	5 – 60
10	25.4	4.660	.310	5 – 60
	35.0	4.944	.594	5 – 60
12	31.8	5.000	.350	5 – 60
	35.0	5.078	.428	5 – 60
	40.8	5.250	.460	5 – 60
	50.0	5.477	.687	5 – 60
15	42.9	5.500	.410	5 – 60
	50.0	5.640	.550	5 – 60
18	54.7	6.000	.460	5 – 60
20	65.4	6.250	.500	5 – 60
24	79.9	7.000	.500	5 – 60

BLACKTOP MIX

- One bag = 66 lbs.
- Coverage at 1/4 in. thick—24 sq. ft.
- Coverage at 1/2 in. thick—12 sq. ft.

BOLTS—CARRIAGE

Length of Bolt (In.)	1/4 In. Dia. Weight/ 1000	5/16 In. Dia. Weight/ 1000	3/8 In. Dia. Weight/ 1000	7/16 In. Dia. Weight/ 1000	1/2 In. Dia. Weight/ 1000	5/8 In. Dia. Weight/ 1000	3/4 In. Dia. Weight/ 1000
1/2	18	39	66	. . .	. . .	. . .	. . .
5/8	20	41	70	. . .	. . .	. . .	. . .
3/4	22	44	74	. . .	. . .	. . .	. . .
7/8	24	46	. . .	. . .	. . .	. . .	. . .
1	25	49	81	115	132	234	388
1 1/8	27	51	. . .	. . .	. . .	. . .	. . .
1 1/4	28	54	88	125	145	255	417
1 1/2	32	59	95	135	158	275	447
1 3/4	35	64	102	145	172	295	476
2	39	69	109	155	185	316	506
2 1/4	43	74	116	165	198	337	535
2 1/2	46	79	124	175	212	357	565
2 3/4	49	84	131	186	225	377	594
3	53	89	138	196	239	398	624
3 1/4	57	94	145	206	252	419	653
3 1/2	60	99	152	216	265	439	683
3 3/4	63	104	159	226	279	459	712
4	67	109	166	236	292	480	742
4 1/4	71	114	174	246	305	501	771
4 1/2	74	119	181	256	319	521	801
4 3/4	78	124	188	266	332	542	. . .
5	81	129	195	276	346	562	860
5 1/2	88	140	209	297	372	603	919
6	95	150	223	317	399	644	978
6 1/2	102	160	237	337	426	685	1037
7	109	170	252	357	453	726	1096
7 1/2	116	180	266	377	479	767	1155
8	123	190	280	398	506	808	1214
8 1/2	. . .	200	295	418	533	849	1273
9	. . .	210	309	438	560	890	1332
9 1/2	. . .	220	323	458	586	931	1391
10	. . .	230	337	478	613	972	1450
10 1/2	. . .	. . .	352	499	640	1013	1509
11	. . .	. . .	366	519	667	1054	1568
11 1/2	. . .	. . .	. . .	539	694	1095	1627
12	. . .	. . .	. . .	559	720	1136	1686

BOLTS—ANCHOR, SILL TO CONCRETE
(TYPICAL SIZES, WEIGHTS)

Size (In.)	Lbs./ 100
$1/2 \times 6$	42
$1/2 \times 8$	52
$1/2 \times 10$	64
$1/2 \times 12$	74
$1/2 \times 14$	82
$1/2 \times 16$	92
$1/2 \times 18$	102

BOLTS—EXPANSION (ALLOY)

Bolt Size (In.)	Length Shield (In.)	Drill Size (In.)
$1/4$	$1 3/8$	$1/2$
$5/16$	$1 3/4$	$5/8$
$3/8$	$2 3/8$	$3/4$
$1/2$	$2 7/8$	$7/8$
Safe loads 200 to 1000 lbs.		

Numerous types and sizes available

BOLTS—EXPANSION (IRON)

Bolt Size (In.)	Length Shield (In.)	Drill Size (In.)
$1/4$	$1 5/8$	$1/2$
$5/16$	$1 7/8$	$5/8$
$3/8$	$2 1/8$	$3/4$
$1/2$	$2 3/8$	$7/8$
$5/8$	$2 7/8$	1
$3/4$	4	$1 1/4$
$7/8$	$4 1/2$	$1 1/2$
1	$4 7/8$	$1 3/4$
Safe loads 200 to 2000 lbs.		

Numerous types and sizes available

BOLTS—MACHINE

Size (In.)	Weight (Lbs.)/100
$3/8 \times 3 1/2$	15
$3/8 \times 4$	16
$3/8 \times 4 1/2$	18
$3/8 \times 5$	19
$1/2 \times 4$	32
$1/2 \times 4 1/2$	35
$1/2 \times 5$	37
$1/2 \times 6$	43
$1/2 \times 7$	48
$1/2 \times 8$	53
$1/2 \times 9$	58
$1/2 \times 10$	64
$1/2 \times 11$	69
$1/2 \times 12$	74
$1/2 \times 13$	80
$1/2 \times 14$	85
$1/2 \times 15$	90
$1/2 \times 16$	96
$5/8 \times 8$	86
$5/8 \times 9$	94
$5/8 \times 10$	102
$5/8 \times 11$	110
$5/8 \times 12$	118
$5/8 \times 13$	126
$5/8 \times 14$	134
$5/8 \times 15$	142
$5/8 \times 16$	150
$5/8 \times 17$	158
$5/8 \times 18$	166
$5/8 \times 20$	182
$5/8 \times 22$	198
$5/8 \times 24$	214
$5/8 \times 26$	230

BOLTS—TOGGLES

Bolt Size (In.)	Bolt Lengths (In.)	Drill Size (In.)
$1/8$	2 – 4	$3/8$
$3/16$	2 – 6	$1/2$
$1/4$	3 – 6	$5/8$
$5/16$	3 – 6	$7/8$
$3/8$	3 – 6	$7/8$
$1/2$	4 – 6	$1 – 1 1/8$

BRIDGING—STEEL

Length (In.)	Size	Joist Centers (In.)	Packed Carton	Wt. (Lbs.)/ Carton
17	1″ × 18 ga.	12	200 pcs. 100 sets	56
20	1″ × 18 ga.	16	200 pcs. 100 sets	80

BRIDGING—WOOD
Pieces of Bridging Required/100 Square Feet Floor

- Joists up to 12 ft. long:
  12-in. centers—20 pcs., 30 lin. ft.
  16-in. centers—16 pcs., 24 lin. ft.

- Joists up to 20 ft. long:
  12-in. centers—40 pcs., 60 lin. ft.
  16-in. centers—32 pcs., 48 lin. ft.

BULLETIN BOARDS—CORK

Cork Thickness (In.)	Mounting	Total Thickness (In.)	Max. Size (Ft.)	Wt. (Lbs.) /Sq. Ft.
1/8	1/4″ Hardboard	3/8	4 × 12	1 1/2
1/8	3/8″ Ins. Board	1/2	4 × 12	1 1/2
1/4	None	1/4	5 × 75	5/8
1/4	1/8″ Hardboard	3/8	4 × 12	1 1/2
1/4	1/4″ Hardboard	1/2	4 × 12	2
1/4	None	1/4	6 × 85	1
1/4	1/8″ Hardboard	3/8	4 × 12	1 3/4
1/4	1/4″ Hardboard	1/2	4 × 12	2 1/4

Many sizes and types available

CALCIUM CHLORIDE

- Packed in: 25 – 100 pound bags
- For dust laying: use ½ pound/square yard for 1 application

CALCIUM CHLORIDE—TO PREVENT FREEZING OF LIQUID-FILLED TRACTOR TIRES

Tire Size	To: −20° Calcium Chl. (Lbs.)	Water (Gal.)	To: −40° Calcium Chl. (Lbs.)	Water (Gal.)
Base	22.8	11.0	36.2	10.2
7.50-24	24.6	11.9	39.1	11.0
-36	34.2	16.5	54.3	15.3
8.25-36	37.6	18.2	59.7	16.8
-40	43.3	20.9	68.8	19.4
9.00-24	34.2	16.5	54.3	15.3
-28	37.6	18.2	59.7	16.8
-36	47.9	23.1	76.0	21.4
-40	51.3	24.7	81.4	23.0
10.00-28	59.3	28.6	94.2	26.5
-36	70.6	34.1	112.2	31.6
-44	82.0	39.6	130.4	36.7
11.25-24	60.4	29.1	95.9	27.0
-28	68.4	33.0	108.6	30.6
-36	82.0	39.6	130.4	36.7
-42	93.4	45.1	148.5	41.8
12.75-24	77.5	37.4	123.1	34.7
-28	86.6	41.8	137.5	38.8
-32	95.7	46.2	152.0	42.8
13.50-24	102.6	49.5	162.9	45.9
-28	114.0	55.0	181.0	51.0
-32	125.4	60.5	199.0	56.0

CALCIUM CHLORIDE—TO PREVENT
WATER FROM FREEZING

Lbs./ Gal. Water	Spec. Grav. at 60°F	Freezing Point F
2.22	1.15	10
2.73	1.18	Zero
3.30	1.21	−10°
3.89	1.24	−20°
4.30	1.26	−30°
4.77	1.28	−40°

CALCIUM CHLORIDE—QUANTITY REQUIRED
FOR DUST LAYING
34% Liquid Solution

Width of Road (Ft.)	Sq. Yds./ Mile	Gal./ Sq. Yd.	Gal./ 100 Lin. Ft.	Gal./ Mile
9	5280	0.15	15	792
		0.20	20	1056
		0.25	25	1320
12	7040	0.15	20	1056
		0.20	27	1408
		0.25	33	1760
14	8213	0.15	23	1232
		0.20	31	1643
		0.25	39	2053
16	9387	0.15	27	1408
		0.20	36	1877
		0.25	44	2347
18	10560	0.15	30	1584
		0.20	40	2112
		0.25	50	2640
20	11733	0.15	33	1760
		0.20	44	2347
		0.25	56	2933
22	12907	0.15	37	1936
		0.20	49	2581
		0.25	61	3227
24	14080	0.15	40	2112
		0.20	53	2816
		0.25	67	3520

CAULKING COMPOUND—QUANTITY REQUIRED

Weight/Gallon: 15 Pounds

1 gallon (231 cu. ft.) will caulk:

120 lin. ft., 3/8-in. joint
75 lin. ft., 1/2-in. joint
30 lin. ft., 3/4-in. joint

1/2 × 1/2-in. ribbon: 77 lin. ft./gallon

CEMENT—KEENE'S

Use—hard waterproof inside finished surface.

- A bag of Keene's cement weighs 100 lbs.
- A bag of Keene's cement contains 1.3 cu.ft.
- A cu. ft. of Keene's cement weighs 75 lbs.
- A 12-qt. bucket holds 30 lbs. of Keene's cement.
- A 12-qt. bucket holds .4 cu. ft. of Keene's cement.

Minimum Mix—1 sack with 25 lbs. dry hydrated lime for hard finish.

CEMENT—MORTAR

Use—all masonry type construction.

- Bag wts.—65 to 75 lbs.
- Bag size—1 cu. ft.

Quantity—1 part mortar, 2 to 3 parts masons sand.

CEMENT—PORTLAND (STRENGTHS BY TYPE)

Compressive Strengths, Minimum P.S.I.
Normal Portland Equals 100%

Types	1 Day	3 Days	7 Days	28 Days
I Normal	-	64	100	143
IA Normal AE	-	52	80	114
II Moderate	-	54	89	143
IIA Moderate AE	-	43	71	114
III High Early	64	125	-	-
IIIA High Early AE	52	100	-	-
IV Low Heat	-	-	36	89
V Sulfate Res.		43	79	107

Strengths and properties of concrete using various types of cement vary. Mixers, mixing time, placing, aggregates, slump (water content), ad mixtures, vibration, climatic conditions, and curing methods all control strengths.

Ad mixtures may control setting time, density, early strength, antifreezing, waterproofing, and other important physical properties.

Specifications are provided on work requiring special concrete strengths and qualities. Suppliers may furnish data and information required.

CEMENT—PORTLAND BUILDING

Weight/Sack: 94 Pounds

I   — For general use.
IA   — For general use. AE added.
II   — Moderate heat of hydration and sulfate action.
IIA — Moderate heat of hydration and sulfate action AE added.
III — High early strength.
IIIA — High early strength. AE added.
IV   — Low heat of hydration.
IVA — Low heat of hydration. AE added.
V   — High sulfate resistance.
VA   — High sulfate resistance. AE added.

CEMENT—WHITE (TYPES)

94-Pound Bags

- White Type I
- White Waterproofed
- White Hi-Early Type III

CEMENT—WHITE MASONRY

- Package: 65-lb. bags, 1 cu. ft.
- Use: 1 bag per 2 to 3 parts white sand.

CHALKBOARDS—SIZES, WEIGHTS

- Thickness—3/16 in.
- Width—3 × 3 1/2 × 4 ft.
- Lengths—5, 6, 7, and 8 ft.
- Weight—2 1/2 lbs./sq.ft.
- Colors—black or green

Other sizes available

CHIMNEYS—MATERIALS REQUIRED

Materials for each Foot of Chimney Height (1/2-Inch Joints)

Material	8 × 8 Modular		8 × 12 Modular		12 × 12 Modular	
	1 Flue	2 Flues	1 Flue	2 Flues	1 Flue	2 Flues
Flue liner	1 ft.	2 ft.	1 ft.	2 ft.	1 ft.	2 ft.
Brick	28	46	32	60	37	70
Cu. ft. mortar	1/2	1	3/4	1 1/4	1	1 1/2

CHIMNEY SCREENS—SIZES, WEIGHTS

Flue Sizes		Wt. (Lbs.)
8 1/2 × 8 1/2	8 × 8	8
8 1/2 × 13	8 × 12	9
13 × 13	12 × 12	10
13 × 18		12
	16 × 16	12
18 × 18		13
20 × 20	20 × 20	16
24 × 24	24 × 24	18

CLOSET LINING—CEDAR

3/8 Inch Thick, 40 Board Feet/Bundle

Size (In.)	Figure as:	Waste (%)
2	2 1/2	25
2 1/2	3	20
3	3 3/4	25
3 1/4	4	23
3 1/2	4 1/4	21
4	4 3/4	19

16

COLD ROOMS—TEMPERATURE REQUIRED

Product	Recom. Temp. (°F)		Product	Recom. Temp. (°F)	
Apples	30 to	32	Maple Syrup and Sugar	40 to	45
Asparagus		32	Meats (canned)	30 to	40
Avocados	40 to	55	Meats (brined)	35 to	43
Bananas (holding)		56	Meats (fresh)	33 to	35
Beef (fresh)	32 to	34	Melons	35 to	40
Beer (barrels)		45	Milk	32 to	38
Berries (fresh, 10 days)	31 to	32	Nursery Stock		30
Butter (short carry)		35	Nuts (in shells)	30 to	40
Butter (long carry)	–10 to	0	Oleomargarine	–10 to	0
Cabbage		32	Onions		32
Cantaloupes	32 to	34	Oranges	32 to	34
Carrots		32	Oysters (tubs)	25 to	35
Celery	31 to	32	Oysters (shells)	33 to	43
Cheese	approx.	34	Peaches	31 to	32
Chocolate (dipping room)		65	Pears	29 to	31
Cream (short carry)	20 to	25	Pork (fresh)	32 to	34
Cream (long carry)	0 to	20	Potatoes	38 to	50
Eggs	30 to	35	Quick-frozen foods - to freeze	–15 to	–40
Fish (fresh)	33 to	40	Quick-frozen foods - to store	–10 to	0
Fruits (dried)	32 to	50	Raisins	40 to	45
Furs (dressed)	25 to	35	Salt Meat - Curing Room		32
Grapes	30 to	32	Sauerkraut	35 to	38
Hams (not brined)	20 to	35	Sausage Casings	40 to	45
Honey	36 to	45	Scallops (frozen)		16
Ice		28	Tobacco	35 to	42
Ice Cream		15	Wines	40 to	50
Lard	32 to	33	Woolens	25 to	28
Lemons	55 to	58			
Livers	20 to	30			

COLUMNS—ADJUSTABLE STEEL (SIZES, WEIGHTS)

Lengths	11 Ga. Weights, ea. (Lbs.)	
	3″ O.D.	4″ O.D.
6′ to 6′ – 3″	29	37
6′ – 3″ to 6′ – 6″	30	38
6′ – 6″ to 6′ – 9″	31	39
6′ – 9″ to 7′	32	40
7′ to 7′ – 3″	33	41
7′ – 3″ to 7′ – 6″	34	42
7′ – 6″ to 7′ – 9″	35	43
7′ – 9″ to 8′	36	44

COLUMNS—ALUMINUM FLUTED (SIZES, TYPES)

Type	Diameters (In.)	Lengths (Ft.)
Round	5, 6, 8, 10, 12, 15, 18, 24	8, 9, 10 up to 30 (in 2-ft. increments)
Square	4×4, 6×6, 8×8, 10×10, 12×12	8, 9, 10 up to 30 (in 2-ft. increments)
Plantation	6×6	8

Available with caps and bases, in various types.
Finish is baked on. Shipped knocked down.

COLUMNS—HOLLOW ROUND STEEL
Cubic Contents

Inside Dia. (In.)	Cu. Inches/ Lin. Ft.	Inside Dia. (In.)	Cu. Feet/ Lin. Ft.
3	84	14	1.06
4	151	15	1.22
5	235	16	1.40
6	339	17	1.58
7	462	18	1.77
8	603	19	1.97
9	763	20	2.20
10	848	21	2.40
11	1086	22	2.64
12	1357	23	2.89
13	1592	24	3.14
		25	3.40
		26	3.69
		27	3.98
		28	4.28
		29	4.59
		30	4.91
		31	5.24
		32	5.58
		33	5.94
		34	6.30
		36	7.07
		42	9.62
		48	12.57
		54	15.90
		60	19.63

COLUMNS—PORCH (WOOD)

Type	Size (In.)	Height (Ft.)	Weight (Lbs.)
Round	6	8	40
Round	6	9	45
Square	6	8	30
Square	6	9	35

COLUMNS—SQUARE STEEL
3 1/2" × 3 1/2" × 1/8" — 3/16" 4" × 6" Caps
6" × 6" Bases

Size	Weight each (Lbs.)
6′	36
6′ - 6″	38 1/2
6′ - 8″	39 1/2
6′ - 10″	40 1/2
7′	41 1/2
7′- 6″	44
8′	47
8′ - 6″	49 1/2
9′	52 1/2
9′ - 6″	55
10′	57 1/2

COLUMNS—WOOD
(CLASSICAL)

Size	Weight each (Lbs.)
8″ × 8′	75
8″ × 9′	86
10″ × 8′	94
10″ × 9′	108
10″ × 10′	120
12″ × 8′	112
12″ × 9′	128
12″ × 10′	144
14″ × 9′	190
14″ × 10′	205
14″ × 12′	270
16″ × 10′	255
16″ × 12′	295
16″ × 14′	340
16″ × 16′	485
18″ × 18′	425
18″ × 20′	470
20″ × 18′	500
20″ × 20′	550

CONCRETE—AGGREGATE
Maximum Sizes Recommended for Various Types of Construction

Minimum Dimension of Section (In.)	Maximum Size of Aggregate[1] (In.) for—		
	Reinforced Walls, Beams, and Columns	Heavily Reinforced Slabs	Lightly Reinforced or Unreinforced Slabs
5 or less.............		$3/4 - 1^1/2$	$3/4 - 1^1/2$
6 to 11	$3/4 - 1^1/2$	$1^1/2$	$1^1/2 - 3$
12 – 29	$1^1/2 - 3$	3	3 to 6
30 or more	$1^1/2 - 3$	3	6

[1]Based on square screen openings.

CONCRETE—AGGREGATE
Weight in Tons/Cubic Yard

Kind of Aggregate	Compacted	Loose
Sand, dry...........................	1.40 – 1.55	1.30 – 1.45
Sand, moist..........................	1.20 – 1.45	1.05 – 1.35
Coarse aggregate (separated):		
$3/16$ to $3/4$ inch	1.35 – 1.45	1.25 – 1.35
$3/4$ to $1^1/2$ inches	1.30 – 1.40	1.25 – 1.35
$1^1/2$ to 3 inches	1.25 – 1.40	1.20 – 1.35
3 to 6 inches	1.20 – 1.35	1.15 – 1.30
Coarse aggregate (combined):		
$3/16$ to $1^1/2$ inches	1.35 – 1.55	1.30 – 1.45
$3/16$ to 3 inches	1.40 – 1.60	1.30 – 1.55
$3/16$ to 6 inches	1.45 – 1.70	1.35 – 1.60
Sand and gravel combined, dry	1.60 – 1.85	1.50 – 1.75

CONCRETE—AIR & WATER CONTENT

Per Cubic Yard of Concrete and the Proportions of Fine and Coarse Aggregate.
(For Concrete Containing Natural Sand with an F.M. of 2.75 and Average Coarse Aggregate, and Having a Slump of 3 to 4 Inches at the Mixer.)

Max. Size of Coarse Aggregate (In.)	Recommended Air Content (Percent)	*Sand, Percent of Total Aggregate by Solid Volume	Percent Dry-rodded Unit Weight of Coarse Aggregate Unit Volume of Concrete	Air Entrained Concrete Average Water Content (Lb./Yd.^{3})	Air Entrained Concrete with WRA—Average Water Content (Lb./Yd.^{3})
3/8	8	60	41	320	300
1/2	7	50	52	305	285
3/4	6	42	62	280	265
1	5	37	67	265	250
1 1/2	4.5	34	73	245	230
2	4	30	76	230	215
3	3.5	28	81	205	190
6	3	24	87	165	155

*When WRA is used in concrete the sand content should be increased 1 or 2 percent to allow for the loss in mortar volume due to water reduction.

CONCRETE—COMPRESSIVE STRENGTH
Probable Minimum Average Compressive Strength of Concrete for Various Water-Cement Ratios (Pounds/Square Inch)

Water-Cement Ratio by Weight	Compressive Strength at 28 Days	
	Air-entrained Concrete	Air-entrained Concrete with WRA
0.40	5,700	6,500
0.45	4,900	5,600
0.50	4,200	4,800
0.55	3,600	4,200
0.60	3,100	3,600
0.65	2,600	3,100
0.70	2,200	2,700

CONCRETE—COVERAGE
PER CUBIC YARD

Thickness (In.)	Sq. Ft.
1	324
1 1/2	216
2	162
2 1/2	130
3	108
3 1/2	93
4	81
4 1/2	72
5	65
5 1/2	59
6	54
6 1/2	50
7	46
7 1/2	43
8	40
8 1/2	38
9	36
10	32
11	29 1/2
12	27

CONCRETE—CURING
Quantities per Sack of Portland Cement

1 lb.	for temp. above 90°F
1 to 1 1/2 lbs.	for temp. 80 to 90°F
2 lbs.	for temp. 32 to 80°F
2 to 4 lbs.	for temp. below 32°F

CONCRETE—EXPANSION

When slab length or width in feet is:

15	20	25	30	33	40	50	60	75	90	100

Estimated expansion in inches when approximate temperature rise above setting temperature is 60° F:

.05	.07	.09	.11	.12	.14	.18	.22	.27	.32	.36

CONCRETE—LIGHTWEIGHT FLOOR & ROOF FILL

Proportions by Loose Dry Volume	Strength 28 Days (psi)	Approx. Weight Dry (Lb./Cu.Ft.)	Heat Conductivity (k) Btu
1-3-3	1000	85	1.80
1-4-4	600	80	1.70
1-5-5	300	75	1.60
1-0-7	300	60	1.50
1-0-8	225	56	1.45
1-0-10	150	46	1.05

CONCRETE—LIGHTWEIGHT STRUCTURAL
(EXPANDED SHALE AGGREGATE)

Lt. Wt. Concrete		Mixes		Aggregate Required/Cu. Yd.			
Strength (psi)	Approx. Weight (Lbs./ Cu.Ft.)	Mix Loose Volume	Water Gallons per Sack	Cement Sacks	Fine (Cu. Yd.)	Coarse (Cu. Yd.)	Sand (Cu. Yd.)
2000	100	1 - 2¹/₄ - 2¹/₂	9.4	6.4	.53	.60	.17
2500	102	1 - 2 - 2¹/₄	8.2	7.2	.53	.60	.20
3000	104	1 - 1³/₄ - 2	7.3	8.1	.54	.60	.15
3500	106	1 - 1¹/₂ -	6.5	8.8	.49	.65	.17
4000	108	1 - 1¹/₂ - 1³/₄	5.9	9.6	.53	.62	.18

CONCRETE—MATERIALS REQUIRED/CUBIC YARD
Based on 1-2-3, 6-Sack Mix

Material	Cubic Yards								
	¹/₄	¹/₂	1	1 ¹/₂	2	2 ¹/₂	3	3 ¹/₂	4
Portland Cement (lbs.)	141	282	564	1128	1410	1692	1974	2256	2256
Fine Aggregate (lbs.)	360	720	1440	2160	2870	3600	4320	5040	5760
Coarse Aggregate (lbs.)	445	890	1780	2680	3560	4450	5340	6220	7100

CONCRETE—MIXES FOR SMALL JOBS

May Be Used without Adjustment

Maximum Size of Aggregate (In.)[1]	Mix Designation	Approximate Bags Cement/ Cubic Yard of Concrete	Pounds of Aggregate/1-Bag Batch		
			Sand[2]		Gravel or Crushed Stone
			Air-Entrained Concrete[3]	Concrete without Air	
1/2	A	7.0	235	245	170
	B	6.9	225	235	190
	C	6.8	225	235	205
3/4	A	6.6	225	235	225
	B	6.4	225	235	245
	C	6.3	215	225	265
1	A	6.4	225	235	245
	B	6.2	215	225	275
	C	6.1	205	215	290
1 1/2	A	6.0	225	235	290
	B	5.8	215	225	320
	C	5.7	205	215	345
2	A	5.7	225	235	330
	B	5.6	215	225	360
	C	5.4	205	215	380

[1]Procedure: Select the proper maximum size of aggregate. Then, using mix B, add just enough water to produce sufficiently workable consistency. If the concrete appears to be undersanded, use mix A; and if it appears to be oversanded, use mix C.

[2]Weights are for dry sand. If damp sand is used, increase the weight of sand 10 pounds for a 1-bag batch, and if very wet sand is used, add 20 pounds for a 1-bag batch.

[3]Air-trained concrete is specified for all Bureau of Reclamation work. In general, air-entrained concrete should be used in all structures that will be exposed to alternate cycles of freezing and thawing.

CONCRETE—PACKAGED MIX

Concrete Req.'d (Cu. Ft.)	90-Lb. Bags
1	1 1/2
2	3
4	6
6	9
8	12
10	15

CONCRETE—PERLITE

Materials Required for One Cubic Yard of Perlite Concrete

Concrete (Sacks)	Perlite (Cu. Ft.)	Water (Gal.)	Air Entraining Agent (Pints)
6.75	27	61	6 3/4
5.40	27	59 1/2	6 3/4
4.50	27	54	6 3/4
3.85	27	54	6 3/4
3.38	27	54	6 3/4

CONCRETE—PERLITE (TYPICAL MIX DESIGNS)

Dry Concrete Properties							Mix Proportion by Volume				Field Tests
Oven Dry Density (Lb./Cu. Ft.)	Compressive Strength (psi at 28 Days)	Thermal Conductivity "K"	Coefficient of Thermal Expansion (per Unit/°F)	Tensile Strength (psi at 28 Days)	Bond Strength to Steel (psi at 28 Days)	Modulus of Elasticity in Compression (psi at 28 Days)	Cement (Sacks)	Perlite (Cu. Ft.)	Water (Gal./Sack Cement)	Air Entraining Agent (Pints)	Wet Density as Poured (Lb./Cu. Ft.)
36	440	0.77	0.0000061	75	83	248,000	1	4	9	1	50 1/2
30 1/2	270	0.64	0.0000055	50	53	158,000	1	5	11	1 1/4	45 1/2
27	180	0.58	0.0000048	40	23	120,000	1	6	12	1 1/2	40 1/2
24	130	0.54	0.0000045	30	—	94,000	1	7	14	1 3/4	38
22	95	0.51	0.0000043	20	—	69,000	1	8	16	2	36 1/2

CONCRETE—PRECAST HOLLOW FLOOR & ROOF SLABS
(SIZES, WEIGHTS, SPANS)

Net Size (In.)	Wt. Grouted/ S.F.	Rec. Spans		Superimposed Allowed Loads at Rec. Spans	
		Floors	Roofs	Floor	Roofs
8 × 15 3/8	55	26	30	65	30
6 × 11	45	22	24	40	30
6 5/16 × 11 5/8	45	22	44	40	44
		PRESTRESSED			
8 × 48	48	30	32	65	60
6 × 48 (Solid)	50	–	24	–	55

Specific products will vary

CONCRETE—PRECAST ROOF SLABS

Type	Thickness	Width	Length	Weight (Lbs./ Sq. Ft.)
Acoustical	4 1/2"	2'	8' 4"	
Concrete Lt. Wt.	3 1/2"	2'	8' 4"	15
Nailable	2"	2'	6'	15
	2 3/4"	2'	8'	20
Gypsum	2"	15"	10'	12

Specific products will vary.

CONCRETE—PRESSURE
Relationship of Concrete Temperature, Pouring Rate and Pressure.

Concrete Placement Rate, Feet/ Hour	Approx. Concrete Pressure in Lb./Square Foot at Concrete Temperature of:			
	40°	50°	60°	70°
1	490	440	390	350
2	700	600	520	460
3	900	760	660	590
4	1100	920	800	700
5	1300	1080	930	820
6	1510	1240	1060	930
7	1700	1400	1200	1050
8	1790	1450	1250	1190

CONCRETE—PROPORTIONS REQUIRED

Type Job	Proportion	Aggregate Maximum Size (In.)
Areaways	1:2½:4	1½
Barn Approaches	1:2½:4	1½
Bins	1:2:3	1½
Boiler Settings	1:2:4	2
Catch Basins	1:2:3	1½
Cisterns	1:2:3	1
Cold Frames	1:2½:4	1
Curbs	1:2½:4	1½
Driveways	1:2:4	1½
Engine Bases	1:2:4	2
Fence Posts	1:2:3	½
Floors, Reinforced	1:2:3	1
Floors, Plain	1:2½:4	1½
Foundations (Mass)	1:2½:5	3
Gutters	1:2½:4	1½
Hot Beds	1:2½:4	1
Piers, House	1:2:3½	2
Pools, Swimming	1:2:3	1½
Retaining Walls	1:2:3½	1½
Roads	1:2:3½	2
Roofs	1:2:3	1½
Runways, Airport	1:2½:4	1½
Sidewalks	1:2½:4	1½
Steps and Stairways	1:2½:4	1
Slabs	1:2:3	1½
Septic Tanks	1:2:4	1
Storage Cellar Walls	1:3½:4	1½
Stucco	1:4	¼
Tennis Courts	1:2½:4	1½
Tree Surgery	1:3	¼
Walls	1:2:4	1½
Walls subject to moisture	1:2:3	1½

CONCRETE—QUANTITIES

Materials Needed to Make 1 Cubic Yard

Mix	Cement (Sacks)	Sand (Cu. Yd.)	Gravel (Cu. Yd.)
1:1½:0	15.5	0.86	—
1:2:0	12.8	0.95	—
1:2½:0	11.0	1.02	—
1:3:0	9.6	1.07	—
1:1½:3	7.6	0.42	0.85
1:2:2	8.2	0.60	0.60
1:2:3	7.0	0.52	0.78
1:2:4	6.0	0.44	0.89
1:2½:3½	5.9	0.55	0.77
1:2½:4	5.6	0.52	0.83
1:2½:5	5.0	0.46	0.92
1:3:5	4.6	0.51	0.85
1:3:6	4.2	0.47	0.94

CONCRETE—SIDEWALKS & FLOORS

Materials Required for 100 Square Feet of Sidewalks and Floors

Proportions	1:2:3			1:2:4			1:2½:5			1:3:5		
Thickness (In.)	Cmt. Sacks	Sand (Cu. Yd.)	Large Agg. (Cu. Yd.)	Cmt. Sacks	Sand (Cu. Yd.)	Large Agg. (Cu. Yd.)	Cmt. Sacks	Sand (Cu. Yd.)	Large Agg. (Cu. Yd.)	Cmt. Sacks	Sand (Cu. Yd.)	Large Agg. (Cu. Yd.)
2½	5.4	0.40	0.60	4.6	0.34	0.68	3.9	0.36	0.72	3.4	0.39	0.65
3	6.5	0.48	0.72	5.6	0.41	0.82	4.6	0.43	0.86	4.1	0.47	0.78
3½	7.5	0.56	0.84	6.5	0.48	0.96	5.4	0.50	1.00	4.8	0.55	0.92
4	8.6	0.64	0.95	7.4	0.55	1.10	6.2	0.57	1.14	5.5	0.63	1.05
4½	9.7	0.72	1.07	8.3	0.62	1.23	6.9	0.64	1.28	6.1	0.70	1.17
5	10.8	0.80	1.19	9.3	0.69	1.37	7.7	0.71	1.42	6.8	0.79	1.31
5½	11.8	0.88	1.31	10.2	0.76	1.50	8.5	0.78	1.57	7.5	0.87	1.45
6	12.9	0.96	1.43	11.1	0.82	1.64	9.2	0.86	1.72	8.2	0.94	1.57

CONCRETE—SLUMPS

Recommended Maximum Slumps for Various Types of Concrete Construction

Type of Construction	Maximum Slump in Inches[1]
Heavy mass construction.	2
Canal lining thickness of 3 or more inches[2]	3
Slabs and tunnel inverts	2
Tops of walls, piers, parapets, and curbs	2
Sidewalls and arch in tunnel lining	4
Other structures	3

[1]These maximum slumps are for concrete after it has been deposited, but before it has been consolidated, and are for mixes having air contents as indicated in the table, Concrete—Air & Water Content.

[2]On machine-placed canal and lateral lining, less than 3 inches thick, the slump should be increased to 3 1/2 inches.

CONCRETE—TEMPERATURE & STRENGTHS

Mix Based on Approx. 600 Pounds Cement and 38 Gallons of Water/Cubic Yard

Temperature at Placing and 2 Hrs. Curing Time Degrees F	Per Sq. In. Compressive Strength (Approx.)			
	Days After Placing Concrete			
	7	28	90	180
70	3200	4800	5600	6200
85	3200	4700	5500	5800
100	3200	4600	5400	5600
115	3200	4500	5300	5500

CONCRETE—TOPPING

Materials Required/100 Square Feet

Thickness (In.)	1:2	
	Cement Sacks	Sand Cu. Yd.
1/2	2.0	0.15
3/4	2.9	0.22
1	3.9	0.29
1 1/4	4.9	0.36
1 1/2	5.9	0.43
1 3/4	6.9	0.50
2	7.9	0.58

CONCRETE—VERMICULITE

Mix Proportion	Mtrl's Req'd Mix by Volume Port. Cement (Bags)	Vermiculite Cu. Ft.	Vermiculite Bags	Water (Gallons)	Approx. Density (Lbs./Cu. Ft.)	Compressive Strength (Lbs./Sq. In.)	Thermal Conductivity ("K" in Btu./Hr./In. Thickness)
1:3	1	3	3/4	10	38	380	1.07
1:4	1	4	1	13	30	240	0.79
1:5	1	5	1 1/4	16	27	150	0.69
1:6	1	6	1 1/2	19	24	115	0.65
1:7	1	7	1 3/4	22	23	95	0.62
1:8	1	8	2	26	22	70	0.60
1:9	1	9	2 1/4	29	21	50	0.59
1:10	1	10	2 1/2	32	20	45	0.57
1:16	1	16	4	52	16	11	0.46

CONCRETE—VOLUME
Volume/4 Sacks Cement

Mix	Cement	Sand	Gravel, Stone	Concrete
1 - 2 - 4	4 sks. = .15 cu. yd.	8 cu. ft. = .30 cu. yd.	16 cu. ft. = .60 cu. yd.	18 cu. ft. = .66 cu. yd.

CONCRETE—VOLUME SHRINKAGE
Relation of Volume of Aggregate to Concrete Volume

Mix	Cement	Sand	Coarse Agg.	Concrete Volume
1 - 2 - 4	6.75	13.5	27	30.3
1 - 2 1/2 - 5	5.40	13.5	27	29.
1 - 3 - 6	4.50	13.5	27	28.1

CONCRETE—WATER REDUCTION RATIO

Water Reduction in Gallons/Sack Cement. Based on Air Entrainment

Cement Factor Sks./Cu.Yd.	4	4.5	5	5.5	6	6.5
2% Entr. Air	3 1/2	3	2 1/2	2	1 3/4	1 1/2
3% Entr. Air	5	4	3 1/4	2 3/4	2 1/4	2
4% Entr. Air	5 3/4	5	4	3 1/2	3	2 1/4
5% Entr. Air	7	5 3/4	4 1/2	3 3/4	3 1/4	2 1/2
6% Entr. Air	8	6 1/2	5 1/4	4 1/4	3 3/4	3
7% Entr. Air	8 1/2	7	5 3/4	4 3/4	4	3 1/4
8% Entr. Air	9	7 1/2	6 1/4	5 1/4	4 1/4	3 1/2

CONCRETE—WATER REQUIRED

Proportions			Water Req'd/ Bag		Water Req'd/ Cu.Yd.	
Parts Cement	Parts Sand	Parts Large Agg.	Min. (Gal.)	Max. (Gal.)	Min. (Gal.)	Max. (Gal.)
1	1 1/2	3	5 1/2	6	42	46
1	2	3	5 3/4	6 1/4	40	43 1/2
1	2	4	6	6 1/2	36	39
1	2 1/2	5	7 1/4	7 3/4	36	38 1/2
1	3	6	8 1/4	8 3/4	35	37 1/2

CONCRETE—YIELD

Cubic-Foot (or Unit) Yield of Various Concrete Mix Ratios

Cement	Sand	Gravel	Yield
1	1.5	3.0	3.5
1	2.0	2.0	3.4
1	2.0	3.0	3.9
1	2.0	4.0	4.5
1	2.5	3.5	4.6
1	2.5	4.0	4.8
1	2.5	5.0	5.4
1	3.0	5.0	5.8
1	3.0	6.0	6.4

CONCRETE COLORS—QUANTITY REQUIRED

To Get	Use (Lbs.)/Sk. Cement)
Red	5 to 10 cement colors
Brown	5 cement colors
Buff	5 to 10 cement colors
Blue	5 cement colors
Green	5 cement colors
Black	5 cement colors
Grey	5 cement colors

CONCRETE CURING COMPOUND

- 1 gallon covers 200 to 300 square feet
- Packaged in: 5-, 30-, and 35-gallon containers

CONCRETE FORM COATING

- Packaged: 1- and 5-gallon cans
- Coverage: 1 gallon coats approx. 200 sq. ft.

CONDUCTORS—COPPER (SIZES)

Size AWG MCM	Area Cir. Mils.	Concentric Lay Stranded Conductors		Bare Conductors	
		No. Wires	Diam. Each Wire (In.)	Diam. (In.)	Area (Sq. In.)
18	1620	Solid	.0403	.0403	.0013
16	2580	Solid	.0508	.0508	.0020
14	4110	Solid	.0641	.0641	.0032
12	6530	Solid	.0808	.0808	.0051
10	10380	Solid	.1019	.1019	.0081
8	16510	Solid	.1285	.1285	.0130
6	26240	7	.0612	.184	.027
4	41740	7	.0772	.232	.042
3	52620	7	.0867	.260	.053
2	66360	7	.0974	.292	.067
1	83690	19	.0664	.332	.087
0	105600	19	.0745	.372	.109
00	133100	19	.0837	.418	.137
000	167800	19	.0940	.470	.173
0000	211600	19	.1055	.528	.219
250	250000	37	.0822	.575	.260
300	300000	37	.0900	.630	.312
350	350000	37	.0973	.681	.364
400	400000	37	.1040	.728	.416
500	500000	37	.1162	.813	.519
600	600000	61	.0992	.893	.626
700	700000	61	.1071	.964	.730
750	750000	61	.1109	.998	.782
800	800000	61	.1145	1.030	.833
900	900000	61	.1215	1.090	.933
1000	1000000	61	.1280	1.150	1.039
1250	1250000	91	.1172	1.289	1.305
1500	1500000	91	.1284	1.410	1.561
1750	1750000	127	.1174	1.526	1.829
2000	2000000	127	.1255	1.630	2.087

CONNECTORS—METAL BUILDING CLIPS
(Light Woodframe Construction)

Clip Types	Uses
• Anchor	Anchors various types and sizes of wood to concrete masonry construction.
• Beam	Secures top of wood post or column to wood beam.
• Bridging	For use between floor or ceiling joists as bridging.
• Drywall	Eliminates use of studs at corners and backup strips between panels.
• Framing	For two and three way ties between framing members.
• Jamb	For prehung door installations.
• Joist	Attaches floor and ceiling joists to headers.
• Panel	For joining plywood panels.
• Plate	Connects walls and partitions.
• Post Base	Attaches wood posts and columns to concrete.
• Purlin	To place purlins between rafters and roof trusses.
• Storm	Anchors roof trusses and rafters to walls.

COPPER-ROLLS (SIZES, WEIGHTS)

Widths (In.)	Gauge	Lengths	Wt. per Roll (Lbs.)
6 – 20	16 oz.	Various	80-90-100

COPPER—(SHEETS)

Oz./ Sq. Ft.	Widths (In.)	Lengths (In.)	Wt./ Sheet (Lbs.)
16	24, 30 & 36	96	16, 20 & 24
16	24, 30 & 36	120	20, 25 & 30
20	24, 30 & 36	96	20, 25 & 30
24	24, 30 & 36	96	24, 30 & 36
32	24, 30 & 36	96	32, 40 & 48
40	30	96	50
48	36	96	72

COPPER—STRIPS (SIZES, WEIGHTS)

Oz./ Sq. Ft.	Width (In.)	Length (In.)	Wt./ Strip (Lbs.)
16	10	96	6 2/3
16	12	96	8
16	14 & 15	96	9 1/3 & 10
16	20	96	13 1/3
16	20	120	16 2/3
20	20	96	16 2/3
24	20	96	20
32	20	96	26 2/3

CURBING—VERTICAL SAWN & SPLIT STONE (SIZES, WEIGHTS)

Top Width (In.)	Depth (In.)	Length (Ft.)	Wt./ Lin. Ft. (Lbs.)
3	17	3	100
4	16	3	100
4	18	3	140
5	16	3	105
5	18	3	150
6	18	6	145
6	20	6	175
7	18	6	160
7	20	6	180

DOMES—PLASTIC CEILING

All Dimensions in Inches—Inside Well Openings

Dimension		Dome Rise	Plastic Thickness
Wood Curb	Concrete Curb		
SQUARE			
14 1/4 × 14 1/4	—	1 1/2	1/16
19 × 19	14 1/4 × 14 1/4	1 1/2	1/16
22 1/4 × 22 1/4	17 1/2 × 17 1/2	4	1/16
30 1/4 × 30 1/4	25 1/2 × 25 1/2	3 1/2	1/16
35 × 35	30 1/4 × 30 1/4	4	1/8
37 × 37	32 1/4 × 32 1/4	4	1/8
39 × 39	34 1/4 × 34 1/4	5	1/8
43 × 43	38 1/4 × 38 1/4	5	1/8
46 1/4 × 46 1/4	41 1/2 × 41 1/2	5 1/4	1/8
55 × 55	50 1/4 × 50 1/4	6	1/8
75 × 75	70 1/4 × 70 1/4	10	1/8
RECTANGULAR			
14 1/4 × 22 1/4	9 1/2 × 17 1/2	2 1/2	1/16
14 1/4 × 46 1/4	9 1/2 × 41 1/2	5 1/4	1/16
22 1/4 × 35	17 1/2 × 30 1/4	5	1/8
22 1/4 × 46 1/4	17 1/2 × 41 1/2	5	1/8
25 1/2 × 89 1/2	20 3/4 × 84 3/4	8	3/16
30 1/4 × 37	25 1/2 × 32 1/4	5	1/8
30 1/4 × 46 1/4	25 1/2 × 41 1/2	5 1/4	1/8
30 1/4 × 58	25 1/2 × 53 1/4	6	1/8
30 1/4 × 69 1/2	25 1/2 × 64 3/4	6	1/8
37 × 75	32 1/4 × 70 1/4	6 1/2	1/8
38 × 59	33 1/4 × 54 1/4	5 1/2	1/8
57 1/2 × 69 1/2	52 3/4 × 64 3/4	8 1/4	3/16
57 1/2 × 89 1/2	52 3/4 × 84 3/4	11	3/16
CIRCULAR			
24" Diameter		3	
31		3 1/2	
43		5	
54		6	
67		7	
79		8 1/2	
91		10	

Numerous sizes and shapes available

DOOR FRAMES—STEEL (TYPICAL SIZES)
6'–8" & 7'–0" FRAMES

Type	Single Swing		Single Swing Pairs		6'–8" Frames		
Frame Gauge	16	16	16	16	18	18	18
Jamb Thickness	3 1/4"	4 1/4" 5 1/4" 6 1/4"	3 1/4"	4 1/4" 5 1/4" 6 1/4"	2 3/4"	3 1/4"	4 1/4" 5 1/4" 6 1/4"
Hollow and Solid Plaster	2'–0"	2'–0"	4'–0"	4'–0"	2'–0"	2'–0"	2'–0"
	2'–4"	2'–4"	4'–8"	4'–8"	2'–4"	2'–4"	2'–4"
	2'–6"	2'–6"	5'–0"	5'–0"	2'–6"	2'–6"	2'–6"
	2'–8"	2'–8"	5'–4"	5'–4"	2'–8"	2'–8"	2'–8"
	3'–0"	3'–0"	6'–0"	6'–0"	3'–0"	3'–0"	3'–0"
Wood Stud	2'–0"	2'–0"	4'–0"	4'–0"		2'–0"	2'–0"
	2'–4"	2'–4"	4'–8"	4'–8"		2'–4"	2'–4"
	2'–6"	2'–6"	5'–0"	5'–0"		2'–6"	2'–6"
	2'–8"	2'–8"	5'–4"	5'–4"		2'–8"	2'–8"
	3'–0"	3'–0"	6'–0"	6'–0"		3'–0"	3'–0"
Masonry		2'–0"		4'–0"			2'–0"
		2'–4"		4'–8"			2'–4"
		2'–6"		5'–0"			2'–6"
		2'–8"		5'–4"			2'–8"
		3'–0"		6'–0"			3'–0"

6' – 8" Frames

Type	Single Swing 18 / 2 3/4"	Single Swing 18 / 3 1/4"	Single Swing 18 / 4 1/4,5 1/4,6 1/4"	Double Swing 18 / 2 3/4"	Double Swing 18 / 3 1/4"	Double Swing 18 / 4 1/4,5 1/4,6 1/4"	Single Swing Pairs 18 / 2 3/4"	Single Swing Pairs 18 / 3 1/4"	Single Swing Pairs 18 / 4 1/4,5 1/4,6 1/4"
Frame Gauge Jamb Thickness									
Hollow and Solid Plaster	1'-6" 2'-0" 2'-4" 2'-6" 2'-8" 3'-0"	1'-6" 2'-0" 2'-4" 2'-6" 2'-8" 3'-0"	1'-6" 2'-0" 2'-4" 2'-6" 2'-8" 3'-0"	2'-6 1/4"	2'-6 1/4"	2'-6 1/4"	4'-0" 4'-8" 5'-0" 5'-4" 6'-0"	4'-0" 4'-8" 5'-0" 5'-4" 6'-0"	4'-0" 4'-8" 5'-0" 5'-4" 6'-0"
Wood Stud		1'-6" 2'-0" 2'-4" 2'-6" 2'-8" 3'-0"	1'-6" 2'-0" 2'-4" 2'-6" 2'-8" 3'-0"		2'-6 1/4"	2'-6 1/4"		4'-0" 4'-8" 5'-0" 5'-4" 6'-0"	4'-0" 4'-8" 5'-0" 5'-4" 6'-0"

Other types, gauges, and sizes available

DOORS—CELLAR BULKHEAD
Steel Sloping Outside

Width	Length	Rise	Steel	Wt. (Lbs.)
3′–11″	4′–10″	2′	12 ga.	160
4′–3″	5′–4″	1′–10″	12 ga.	180
4′–7″	6′–0″	1′–7″	12 ga.	210

Other sizes available

DOORS—FRENCH (SIZES)

1′–6″ × 6′–8″	1 3/8″
2′–0″ × 6′–8″	1 3/8″
2′–6″ × 6′–8″	1 3/8″
2′–8″ × 6′–8″	1 3/8″
3′–0″ × 6′–8″	1 3/8″
2′–0″ × 6′–8″	1 3/4″
2′–6″ × 6′–8″	1 3/4″
3′–0″ × 6′–8″	1 3/4″

Wt. Open—35 lbs.; Glazed—40 lbs.

DOORS—FOLDING FABRIC (SIZES)

Width	Height	Width Closed	Weight (Lbs.)
2′–4″	6′–8 1/2″	6 5/8″	50
2′–10″	6′–8 1/2″	7 5/8″	55
2′–4″	6′–8″	6″	50
2′–10″	6′–8″	7″	55
3′–5″	6′–8″	8″	55
3′–11″	6′–8″	9″	65
4′–5″	6′–8″	10″	70
5′–0″	6′–8″	11″	75
6′–0″	6′–8″	13″	85
7′–1″	6′–8″	15″	95
8′–2″	6′–8″	17″	110
3′–11″	8′–0″	9″	75
4′–5″	8′–0″	10″	85
5′–0″	8′–0″	11″	90
6′–0″	8′–0″	13″	95
7′–1″	8′–0″	15″	105
8′–2″	8′–0″	17″	130
9′–3″	8′–0″	19″	145
10′–3″	8′–0″	21″	170

Other sizes and styles available

DOORS—FOLDING FABRIC
Panel Widths: 4 1/4 Inch

Opening Widths (In.)	Number of Folds	Stack Back (In.)	Weights (Lbs.) for Heights	
			78" or 80"	96"
24	5	4 1/4	11	13
32	6	4 3/4	12	15
36	8	5 11/16	14	17
48	10	6 11/16	16	19
60	14	8 5/8	20	23
72	18	10 1/16	25	29
84	21	11 1/2	30	35
96	24	13	36	40

DOORS—FOLDING WOOD

Opening Width	Stack Size	Opening Width	Stack Size	Opening Width	Stack Size	Opening Width	Stack Size
1' 9"	4"	8' 11 1/4"	12 5/8"	16' 1 1/2"	21 1/4"	23' 3 3/4"	29 7/8"
2' 0 3/4"	4 3/8"	9' 3"	13"	16' 5 1/4"	21 5/8"	23' 7 1/2"	30 1/4"
2' 4 1/2"	4 3/4"	9' 6 3/4"	13 3/8"	16' 9"	22"	23' 11 1/4"	30 5/8"
2' 8 1/4"	5 1/8"	9' 10 1/2"	13 3/4"	17' 0 3/4"	22 3/8"	24' 3"	31"
3' 0"	5 1/2"	10' 2 1/4"	14 1/8"	17' 4 1/2"	22 3/4"	24' 6 3/4"	31 3/8"
3' 3 3/4"	5 7/8"	10' 6"	14 1/2"	17' 8 1/4"	23 1/8"	24' 10 1/2"	31 3/4"
3' 7 1/2"	6 1/4"	10' 9 3/4"	14 7/8"	18' 0"	23 1/2"	25' 2 1/4"	32 1/8"
3' 11 1/4"	6 5/8"	11' 1 1/2"	15 1/4"	18' 3 3/4"	23 7/8"	25' 6"	32 1/2"
4' 3"	7"	11' 5 1/4"	15 5/8"	18' 7 1/2"	24 1/4"	25' 9 3/4"	32 7/8"
4' 6 3/4"	7 3/8"	11' 9"	16"	18' 11 1/4"	24 5/8"	26' 1 1/2"	33 1/4"
4' 10 1/2"	7 3/4"	12' 0 3/4"	16 3/8"	19' 3"	25"	26' 5 1/4"	33 5/8"
5' 2 1/4"	8 1/8"	12' 4 1/2"	16 3/4"	19' 6 3/4"	25 3/8"	26' 9"	34"
5' 6"	8 1/2"	12' 8 1/4"	17 1/8"	19' 10 1/2"	25 3/4"	27' 0 3/4"	34 3/8"
5' 9 3/4"	8 7/8"	13' 0"	17 1/2"	20' 2 1/4"	26 1/8"	27' 4 1/2"	34 3/4"
6' 1 1/2"	9 1/4"	13' 3 3/4"	17 7/8"	20' 6"	26 1/2"	27' 8 1/4"	35 1/8"
6' 5 1/4"	9 5/8"	13' 7 1/2"	18 1/4"	20' 9 3/4"	26 7/8"	28' 0"	35 1/2"
6' 9"	10"	13' 11 1/4"	18 5/8"	21' 1 1/2"	27 1/4"	28' 3 3/4"	35 7/8"
7' 0 3/4"	10 3/8"	14' 3"	19"	21' 5 1/4"	27 5/8"	28' 7 1/2"	36 1/4"
7' 4 1/2"	10 3/4"	14' 6 3/4"	19 3/8"	21' 9"	28"	28' 11 1/4"	36 5/8"
7' 8 1/4"	11 1/8"	14' 10 1/2"	19 3/4"	22' 0 3/4"	28 3/8"	29' 3"	37"
8' 0"	11 1/2"	15' 2 1/4"	20 1/8"	22' 4 1/2"	28 3/4"	29' 6 3/4"	37 3/8"
8' 3 3/4"	11 7/8"	15' 6"	20 1/2"	22' 8 1/4"	29 1/8"	29' 10 1/2"	37 3/4"
8' 7 1/2"	12 1/4"	15' 9 3/4"	20 7/8"	23' 0"	29 1/2"	30' 2 1/4"	38 1/8"

DOORS—STEEL
1 3/4" Size, 6' – 8" Doors

Types	Single Swing	Single Swing Pairs
Flush	2' – 0"	(2) 2' – 0"
	2' – 4"	(2) 2' – 4"
	2' – 6"	(2) 2' – 6"
	2' – 8"	(2) 2' – 8"
	3' – 0"	(2) 3' – 0"
Half Glass	2' – 0"	(2) 2' – 0"
	2' – 4"	(2) 2' – 4"
	2' – 6"	(2) 2' – 6"
	2' – 8"	(2) 2' – 8"
	3' – 0"	(2) 3' – 0"
Vision Panel	2' – 0"	(2) 2' – 0"
	2' – 4"	(2) 2' – 4"
	2' – 6"	(2) 2' – 6"
	2' – 8"	(2) 2' – 8"
	3' – 0"	(2) 3' – 0"
Bottom Louver	2' – 0"	(2) 2' – 0"
	2' – 4"	(2) 2' – 4"
	2' – 6"	(2) 2' – 6"
	2' – 8"	(2) 2' – 8"
	3' – 0"	(2) 3' – 0"
Top Louver	2' – 0"	(2) 2' – 0"
	2' – 4"	(2) 2' – 4"
	2' – 6"	(2) 2' – 6"
	2' – 8"	(2) 2' – 8"
	3' – 0"	(2) 3' – 0"
Half Glass and Louver	2' – 0"	(2) 2' – 0"
	2' – 4"	(2) 2' – 4"
	2' – 6"	(2) 2' – 6"
	2' – 8"	(2) 2' – 8"
	3' – 0"	(2) 3' – 0"
Vision Panel and Louver	2' – 0"	(2) 2' – 0"
	2' – 4"	(2) 2' – 4"
	2' – 6"	(2) 2' – 6"
	2' – 8"	(2) 2' – 8"
	3' – 0"	(2) 3' – 0"
"DL" Double Louver	2' – 0"	(2) 2' – 0"
	2' – 4"	(2) 2' – 4"
	2' – 6"	(2) 2' – 6"
	2' – 8"	(2) 2' – 6"
	3' – 0"	(2) 3' – 0"

Stiles 18 ga., Panels 20 ga.

Other types, gauges and sizes available

DOORS—STEEL
1 3/8″ Size, 6′ – 8″ Doors

Types	Single Swing	Double Swing	Single Swing Pairs	Double Swing Double
Flush	1′ – 6″			
	2′ – 0″		(2) 2′ – 0″	
	2′ – 4″		(2) 2′ – 4″	
	2′ – 6″	2′ – 6″	(2) 2′ – 6″	(2) 2′ 6″
	2′ – 8″		(2) 2′ – 8″	
	3′ – 0″		(2) 3′ – 0″	
Bottom Louver	1′ – 6″			
	2′ – 0″		(2) 2′ – 0″	
	2′ – 4″		(2) 2′ – 4″	
	2′ – 6″	2′ – 6″	(2) 2′ – 6″	(2) 2′ 6″
	2′ – 8″		(2) 2′ – 8″	
	3′ – 0″		(2) 3′ – 0″	
"V" Vision Panel	2′ – 0″		(2) 2′ – 0″	
	2′ – 4″		(2) 2′ – 4″	
	2′ – 6″	2′ – 6″	(2) 2′ – 6″	(2) 2′ 6″
	2′ – 8″		(2) 2′ – 8″	
	3′ – 0″		(2) 3′ – 0″	
Top Louver	1′ – 6″			
	2′ – 0″		(2) 2′ – 0″	
	2′ – 4″		(2) 2′ – 4″	
	2′ – 6″	2′ – 6″	(2) 2′ – 6″	(2) 2′ – 6″
	2′ – 8″		(2) 2′ – 8″	
	3′ – 0″		(2) 3′ – 0″	
Double Louver	1′ – 6″			
	2′ – 0″		(2) 2′ – 0″	
	2′ – 4″		(2) 2′ – 4″	
	2′ – 6″	2′ – 6″	(2) 2′ – 6″	(2) 2′ – 6″
	2′ – 8″		(2) 2′ – 8″	
	3′ – 0″		(2) 3′ – 0″	
Vision Panel and Louver	2′ – 0″		(2) 2′ – 0″	
	2′ – 4″		(2) 2′ – 4″	
	2′ – 6″	2′ – 6″	(2) 2′ – 6″	(2) 2′ – 6″
	2′ – 8″		(2) 2′ – 8″	
	3′ – 0″		(2) 3′ – 0″	

Stiles 18 ga., Panels 20 ga.

Other types, gauges and sizes available

DOORS—WOOD (COMBINATION & SCREEN)

Width	Height
2'–6"	6'–7"
2'–6"	6'–9"
2'–8"	6'–9"
2'–8"	7'–1"
2'–10"	6'–11"
2'–10"	7'–1"
3'–0"	6'–9"
3'–0"	7'–1"

Av. wt. 50 lbs., 1 1/8" thick

DOORS—WOOD (BIFOLD)

- Thickness—1 3/8"
- Height—6'–8 1/2"
- Widths, 2-door—1'–6", 2'–0", 2'–8", 3'–0"
- Widths, 4-door—3'–0", 4'–0", 5'–0", 6'–0", 7'–0", 8'–0"

Available in special sizes.
Available louvered, paneled, or combination.

DOORS—WOOD (LOUVER)

- Thicknesses—1 3/8" and 1 3/4"
- Heights—6'–6", 6'–8", 7'–0"
- Widths—1'–0", 1'–3", 1'–4", 1'–6", 1'–8", 2'–0", 2'–4", 2'–6", 2'–8", 3'–0"

Available with raised-panel bottom sections

DOORS—WOOD (RESIDENTIAL 20-MINUTE FIRE-RATED)

- Thickness - 1 3/4"
- Height - 6'–8"
- Widths - 2'–6", 2'–8", 3'–0"

Special sizes available to 4 by 9 ft.

DOORS—WOOD (SIZES)

Standard Sizes

$1'-6''\times6'-6''$
$1'-6''\times6'-8''$
$2'-0''\times6'-0''$
$2'-0''\times6'-6''$
$2'-0''\times6'-8''$
$2'-0''\times7'-0''$
$2'-4''\times6'-0''$
$2'-4''\times6'-6''$
$2'-4''\times7'-0''$
$2'-6''\times6'-0''$
$2'-6''\times6'-6''$
$2'-6''\times6'-8''$
$2'-6''\times7'-0''$
$2'-8''\times6'-0''$
$2'-8''\times6'-6''$
$2'-8''\times6'-8''$
$2'-8''\times7'-0''$
$3'-0''\times6'-8''$
$3'-0''\times7'-0''$

Average Weights (Lbs.)

$1\,^3/_8''$	Panel (Colonial)	40
$1\,^3/_8''$	Flush	40
$1\,^3/_8''$	French Glazed	40
$1\,^3/_4''$	Panel (Colonial)	50
$1\,^3/_4''$	Flush (Outside)	60

DOORS—WOOD (SCREEN)

Width	Height
$2'-6''$	$6'-7''$
$2'-6''$	$6'-9''$
$2'-8''$	$6'-9''$
$2'-8''$	$7'-1''$
$2'-10''$	$6'-11''$
$2'-10''$	$7'-11''$
$3'-0''$	$6'-9''$
$3'-0''$	$7'-1''$

All $1\,^1/_8''$ thick

DOORS—WOOD (WEIGHTS)
Pounds/Square Foot

Thickness (In.)	1 3/8	1 3/4	2	2 1/4	2 1/2	3
Ash	4 1/2	5 1/4	6	6 3/4	7 1/2	9
Birch	3 3/4	4 1/3	5	5 2/3	6 1/4	7 1/2
Chestnut	3 3/4	4 1/3	5	5 2/3	6 1/4	7 1/2
Fir	3	3 1/2	4	4 1/2	5	6
Mahogany	4 1/2	5 1/4	6	6 3/4	7 1/2	9
Redwood	3 1/4	3 3/4	4 1/3	4 3/4	5 1/2	6 1/2
White Oak	6	7	8	9	10	12
White Pine	3	3 1/2	4	4 1/2	5	6
Yellow Pine	5 1/2	6 1/2	7 1/3	8 1/4	9 1/5	11
Hollow Metal		6 1/2				
Hollow Metal (1/2 glass panel)		5				
Hollow Metal (full glass panel)		4 1/2				

DRAIN TILE—ACRES DRAINED
Average Acres Drained/100 Feet of Pipe/24 Hours

Tile Dia. (In.)	Slope; Fall/100 Linear Feet									
	.08	.10	.15	.20	.25	.30	.40	.50	.75	1.00
4	3	3	4	4	5	5	6	7	9	10
5	5	6	7	9	9	10	12	13	16	19
6	9	10	12	14	15	18	20	22	27	31
7	13	15	19	22	24	27	31	34	42	8
8	20	22	27	32	36	39	45	50	61	71
10	37	41	50	58	65	71	82	92	112	130
12	61	67	83	96	108	118	137	153	187	213
14	95	105	129	149	167	182	211	235	288	333
15	113	127	155	179	200	218	252	283	346	400
16	134	149	183	211	238	260	302	336	413	473
18	187	209	256	295	328	360	417	467	570	660
20	249	280	341	393	440	483	556	625	760	880
22	323	360	442	510	571	626	720	800	987	1,140
24	407	453	558	643	720	794	913	1,020	1,247	1,440

DRY MATERIALS—COVERAGE
Pounds/100 Square Feet

Thickness (In.)	Cinders	Gravel	Slag	Limestone	Dry Fill	Dry Top Soil
1	500	850	650	750	725	650
2	1000	1700	1300	1500	1425	1300
3	1500	2550	1950	2250	2175	1950
4	2000	3400	2600	3000	2850	2600
5	2500	4250	3250	3700	3625	3250
6	3000	5100	3900	4400	4350	3900

DUCT—ROUND FIBER AIR
(SIZES, WEIGHTS)

Inside Dia. (In.)	Wall Thickness (In.)	Weight/1,000 Ft. (Lbs.)
2	.125	278
3	.125	407
3 1/2	.150	562
4	.150	635
4 1/2	.150	712
5	.150	788
6	.175	1,086
6 1/2	.175	1,195
7	.200	1,439
7 1/2	.200	1,538
8	.200	1,636
8 1/2	.200	1,755
9	.225	2,071
10	.225	2,296
11	.225	2,516
12	.225	2,745
14	.250	3,523
16	.300	4,777
18	.300	5,365
20	.375	7,372
22	.375	8,095
24	.375	8,818
26	.375	9,515
28	.375	10,236
30	.400	11,663
32	.400	12,428
36	.410	14,303

Products Vary

E

EQUIVALENTS—CUBIC CONVERSION

Units	Cubic Inches	Cubic Feet	Cubic Yards
1 cubic inch =	1	0.000 578 704	0.000 021 433
1 cubic foot =	1728	1	0.037 037 0
1 cubic yard =	46 656	27	1
1 cubic centi. =	0.061 023 38	0.000 035 314	0.000 001 308
1 cubic deci. =	61.023 38	0.035 314 45	0.001 307 943
1 cubic meter =	61 023.38	35.314 45	1.307 942 8

EQUIVALENTS—METRIC (SOFT)

decimeter . . . . 4 inches
meter . . . . 1.1 yards
kilometer . . . . 5/8 of mile
hectare . . . . 2 1/2 acres
Stere, or cu. meter . 1/4 of a cord
liter . . . . 1.06 qt. or 0.9 qt. dry
hektoliter . . . . 2.8 bushels
kilogram . . . . 2.2 pounds
metric ton . . . . 2,200 pounds

EQUIVALENTS—DECIMAL

1/64	.015625	33/64	.515625
1/32	.03125	17/32	.53125
3/64	.046875	35/64	.546875
1/16	.0625	9/16	.5625
5/64	.078125	37/64	.578125
3/32	.09375	19/32	.59375
7/64	.109375	39/64	.609375
1/8	.125	5/8	.625
9/64	.140625	41/64	.640625
5/32	.15625	21/32	.65625
11/64	.171875	43/64	.671875
3/16	.1875	11/16	.6875
13/64	.203125	45/64	.703125
7/32	.21875	25/32	.71875
15/64	.234375	47/64	.734375
1/4	.25	3/4	.75
17/64	.265625	49/64	.765625
9/32	.28125	25/32	.78125
19/64	.296875	51/64	.796875
5/16	.3125	13/16	.8125
21/64	.328125	53/64	.828125
11/32	.34375	27/32	.84375
23/64	.359375	55/64	.859375
3/8	.375	7/8	.875
25/64	.390625	57/64	.890625
13/32	.40625	29/32	.90625
27/64	.421875	59/64	.921875
7/16	.4375	15/16	.9375
29/64	.453125	61/64	.953125
15/32	.46875	31/32	.96875
31/64	.484375	63/64	.984375
1/2	.5		

EQUIVALENTS—TEMPERATURE CONVERSION

- Freezing Pt. Water—0°C, 32°F
- Boiling Pt. Water—100°C, 212°F

Conversion Equation:
- $t_C = (t_F - 32)/1.8$
- $t_F = (t_C + 32) \times .05556$

EQUIVALENTS—VOLUME & CAPACITY

French	U.S. & British
VOLUME	
1 cubic metre	= 35.314 cu. ft. or 1.308 cu. yds.
0.7645 cubic metre	= 1 cu. yd.
0.02832 cubic metre	= 1 cu. ft.
1 cubic decimetre	= 61.023 cu. ins., or 0.0353 cu. ft.
16.387 cubic centimetres	= 1 cu. in.
1 litre = 1 cubic decimetre	= 61.023 cu. in.
	= 1.05671 qts. U.S.
1 hectolitre or decistere	= 3.5314 cu. ft.
	= 2.8375 bu. U.S.
1 stere, kilolitre, or cubic metre	= 1.308 cu. yds.
	= 28.37 bu. U.S.
CAPACITY	
1 litre = 1 cubic decimetre	= 0.2642 gallon (61.023 cu. ins. (0.03531 cu. ft. (American) (2.202 pounds of water at 62°F)
28.317 litres	= 1 cu. ft.
4.543 litres	= 1 gallon (British)
3.785 litres	= 1 gallon (Amer.)

ESTIMATED TIME REQUIRED— BUILT-UP ROOFING, INSULATION, & FLASHING
Pitch: 1/2 – 3 Inches/Foot

Description	Unit	Man-Hr./Unit
Flashing	1000 lin. ft.	60
Insulation	1000 sq. ft.	25
Roofing	1000 sq. ft.	
2 ply		12
3 ply		20
4 ply		25
5 ply		30
Table includes melting asphalt, laying felt, mopping, and laying gravel.		

ESTIMATED TIME REQUIRED—FINISH CARPENTRY

Description	Unit	Man-Hr./Unit
Baseboard (2 member)	1000 lin. ft.	72
Ceilings	1000 sq. ft.	
cemented tile		32
panel w/ suspension		72
plasterboard (including tape)[1]		64
wood		48
Door frame, trim	ea.	2.5
Installing prefab. closets	ea.	16
Molding (chair)	1000 lin. ft.	48
Plasterboard (complete)	1000 sq. ft.	110
Setting kitchen cabinets	ea.	1.5
Sliding door w/pocket	ea.	8
Shelving	1000 sq. ft.	64
Stairs		
closed stringer, built on job	story	16
closed stringer, prefab.	story	8
open stringer	story	24
Walls	1000 sq. ft.	
plasterboard (including tape)		48
plywood		80
Wood frame, trim	ea.	3

[1]Includes installation of furring strips when necessary.

ESTIMATED TIME REQUIRED—FLOORING

Description	Unit	Man-Hr./Unit
Sheet Vinyl	1000 sq. ft.	32
Soft tile	1000 sq. ft.	
cemented		24
nailed		32
Wood floors	1000 sq. ft.	
Finish floor		
hardwood		32
softwood		24
Subfloor		
plywood		16
tongue & groove		24

ESTIMATED TIME REQUIRED—INSULATION

Description	Unit	Man-Hr./Unit
Acoustic	1000 sq. ft.	
Quilt		8
Strip		24
Thermal	1000 sq. ft.	
Board		
ceiling		24
floor		8
roof		16
wall		32
Foil alone		16
Rigid foam		32
Rock wool/Fiberglass		
batts		24
loose		16

ESTIMATED TIME REQUIRED—METAL & TILE ROOFING
Pitch at Least 3 Inches/Foot

Description	Unit	Man-Hr./Unit
Metal	1000 sq. ft.	
corrugated & V-crimp		
metal purlins		36
wood purlins		18
Tile	1000 sq. ft.	
clay		55
metal		60

Table includes placing, caulking, drilling, and fastening materials.

ESTIMATED TIME REQUIRED—ROUGH FRAMING

Description	Unit	Man-Hr./Unit		
Beams (3-2″ × 8″)	MFBM	40		
Blocking	MFBM	32		
Bridging	100 pairs	5		
Ceiling joists	MFBM	32		
Door bucks	ea.	3		
Floor joists, sills	MFBM	32		
Furring including plugging	1000 lin. ft.	32		
Grounds for plaster	1000 lin. ft.	48		
Rafters	MFBM	48		
Trusses	ea.	Man-Hr. Assembly	Man-Hr. Placement	Hours Hoist Time
Span ft. 20		2.5	4	8
30		5	8	12
40		12	8	16
50		20	6[1]	8[1]
60		24	6[1]	9[1]
80		32	6[1]	11[1]
Wall frames, plates	MFBM	56		

[1]Assumes use of crane.

ESTIMATED TIME REQUIRED—SHEATHING & SIDING

Description	Unit	Man-Hr./Unit
Roof decking	1000 sq. ft.	
plywood		24
tongue & groove		32
Siding	1000 sq. ft.	
corrugated asbestos		32
drop siding		32
narrow bevel		48
plywood		24
shingles		40
Wall sheathing	1000 sq. ft.	
bldg. paper		16
fiberboard		24
tongue & groove		24
plywood		16

ESTIMATED TIME REQUIRED—SHINGLE ROOFING
Pitch at Least 3 Inches/Foot

Description	Unit	Man-Hr./Unit
Asphalt	1000 sq. ft.	30
Metal	1000 sq. ft.	50
Slate	1000 sq. ft.	55
Wood	1000 sq. ft.	35

Table includes placing and nailing.

ESTIMATED TIME REQUIRED—WOOD DOORS

Description	Unit	Man-Hr./Unit
Caulking (w/gun)	1000 lin. ft.	16
Doors w/hardware	ea.	
exterior[1]		2
interior[1]		1.5
manual sliding (including tracks)		8
motorized sliding[2]		56
overhead (including machinery)		16
screendoors		1.5
Weatherstripping	ea. opening	1.5

[1]For double doors add 50% to labor estimates.
[2]Includes tracks and all necessary machinery, with control equipment.

ESTIMATED TIME REQUIRED—WOOD WINDOWS

Description	Unit	Man-Hr./Unit
Caulking (w/gun)	1000 lin. ft.	16
Screens	ea.	1.5
Weatherstripping	ea. opening	1.5
Windows (avg. 20 sq. ft.)	ea.	
casement		1.5
double hung		2.5
jalousie		2.5
louvers		4
skylight		8
sliding		2.5
Venetian blinds	ea.	1

EXCAVATION—BACKFILL (AVERAGE CONDITIONS, POROUS MATERIALS)

Pipe Size (In.)	Trench Width (In.)	Trench Depth (Feet)															
		1	2	3	4	5	6	7	8	9	10	11	12	14	16	18	20
4	12	.04	.07	.11	.15	.18	.22	.26	.30	.33	.37	.41	.44	.52	.59	.67	.74
6	15	.05	.09	.14	.19	.23	.28	.32	.37	.42	.46	.51	.56	.65	.74	.83	.93
8	18	.06	.11	.17	.22	.28	.33	.39	.45	.50	.56	.61	.67	.78	.89	1.00	1.11
10	22	.07	.14	.21	.28	.35	.42	.49	.55	.62	.69	.76	.83	.97	1.11	1.25	1.39
12	27	.08	.17	.25	.33	.42	.50	.58	.67	.75	.83	.92	1.00	1.17	1.33	1.50	1.67
15	33	.10	.20	.30	.41	.51	.61	.71	.81	.92	1.02	1.12	1.22	1.43	1.63	1.83	2.04
18	36	.11	.22	.33	.44	.55	.67	.78	.89	1.00	1.11	1.22	1.33	1.56	1.78	2.00	2.22
21	41	.13	.25	.38	.51	.63	.76	.89	1.01	1.14	1.27	1.39	1.52	1.77	2.03	2.28	2.53
24	46	.14	.28	.43	.57	.71	.85	.99	1.14	1.28	1.42	1.56	1.70	1.99	2.27	2.55	2.84
27	50	.15	.31	.46	.62	.77	.93	1.08	1.24	1.39	1.54	1.70	1.85	2.16	2.47	2.78	3.09
30	53	.16	.33	.49	.66	.82	.98	1.15	1.31	1.47	1.64	1.80	1.96	2.29	2.62	2.94	3.27
33	57	.18	.35	.53	.70	.88	1.05	1.23	1.41	1.58	1.76	1.94	2.11	2.46	2.81	3.17	3.52
36	60	.19	.37	.56	.74	.93	1.11	1.30	1.48	1.67	1.85	2.04	2.22	2.59	2.96	3.33	3.70
42	67	.21	.41	.62	.83	1.03	1.24	1.45	1.65	1.86	2.07	2.27	2.48	2.89	3.31	3.72	4.14
48	74	.23	.46	.69	.91	1.14	1.37	1.60	1.83	2.06	2.28	2.51	2.74	3.20	3.65	4.11	4.57
54	81	.25	.50	.75	1.00	1.25	1.50	1.75	2.00	2.25	2.50	2.75	3.00	3.50	4.00	4.50	5.00
60	88	.27	.54	.82	1.09	1.36	1.63	1.90	2.17	2.44	2.72	2.99	3.26	3.80	4.34	4.89	5.43
66	102	.30	.61	.94	1.26	1.57	1.89	2.20	2.51	2.83	3.14	3.46	3.78	4.41	5.04	5.67	6.30
72	108	.33	.67	1.00	1.33	1.67	2.00	2.33	2.67	3.00	3.33	3.66	4.00	4.67	5.33	6.00	6.67
84	120	.37	.74	1.11	1.48	1.85	2.22	2.59	2.96	3.33	3.70	4.07	4.44	5.19	5.93	6.66	7.41
96	132	.40	.82	1.22	1.63	2.04	2.44	2.85	3.26	3.67	4.07	4.48	4.88	5.70	6.52	7.33	8.14
108	144	.44	.89	1.33	1.78	2.22	2.66	3.11	3.55	4.00	4.44	4.88	5.33	6.22	7.11	8.00	8.96

EXCAVATION—CALCULATIONS

Deep (Ft.)	Per Foot of Depth	
	Sq. Yds.	Cu. Yds.
1	1 yd.	$1/3$
2	1 yd.	$2/3$
3	1 yd.	1
4		$1\,1/3$
5		$1\,1/3$
6		2
7		$2\,1/3$
8		$2\,2/3$
9		3

EXPANDED METALS—SHEET

Size of Diamond (In.)	Gauge	Lbs./ Sq. Ft.	Lengths Long Way of Diamond (Ft.)	Widths Short Way of Diamond
$1\,3/8 \times 3$	17	.20	8	2', 4'
$1\,3/8 \times 3$	17	.40	8	2', 4'
$1/2 \times 1.2$	18	.74	8	3', 4'-6″
$3/4 \times 2$	16	.51	8	4'-4″, 6'-6″
$3/4 \times 2$	12	.80	8	4'-4″, 6'-6″
$3/4 \times 2$	10	1.19	8	3', 4'-6″
$1\,3/8 \times 3$	12	.60	8	3', 4'-6″, 6'
$1\,3/8 \times 3$	12	.80	8	3'-6″, 7'
$1\,3/8 \times 3$	10	1.19	8	3', 4', 6'

Other styles available

EXPANSION—LINEAR, OF MATERIALS

Coefficients of Expansion for 100 Degrees

Substances	Linear Expansion	
	Centigrade	Fahrenheit
METALS AND ALLOYS		
Aluminum, wrought	.00231	.00128
Brass	.00188	.00104
Bronze	.00181	.00101
Copper.	.00168	.00093
Iron, cast, gray.	.00106	.00059
Iron, wrought	.00120	.00067
Iron, wire	.00124	.00069
Lead.	.00286	.00159
Nickel.	.00126	.00070
Steel, cast	.00110	.00061
Steel, hard	.00132	.00073
Steel, medium	.00120	.00067
Steel, soft.	.00110	.00061
Zinc, rolled.	.00311	.00173
STONE AND MASONRY		
Ashlar masonry	.00063	.00035
Brick masonry	.00055	.00031
Cement, portland.	.00107	.00059
Concrete	.00143	.00079
Granite.	.00084	.00047
Limestone	.00080	.00044
Marble	.00100	.00056
Plaster	.00166	.00092
Rubble masonry.	.00063	.00035
Sandstone	.00110	.00061
Slate.	.00104	.00058
TIMBER		
Fir)	.00037	.00021
Maple) Parallel to	.00064	.00036
Oak) fiber	.00049	.00027
Pine)	.00054	.00030
Fir)	.0058	.0032
Maple) Perpendicular	.0048	.0027
Oak) to fiber	.0054	.0030
Pine)	.0034	.0019

EXPANSION JOINT—ASPHALT POURED
Quantity: Pounds/Linear Foot
Weight/Cubic Foot: 65 Pounds
Weight/Gallon: 8 Pounds

Depth of Joint (In.)	Width of Joint (In.)						
	1/4	3/8	1/2	5/8	3/4	7/8	1
1/4	2.8	4.2	5.5	6.9	8.3	9.6	11
1/2	5.5	8.3	11.	13.8	16.5	19.3	22
3/4	8.3	12.4	16.5	20.6	24.8	28.9	33
1	11.	16.5	22.	27.5	33.	38.5	44
1 1/2	16.5	24.8	33.	41.3	49.5	57.8	66
2	22.	33.	44.	55.	66.	77.	88

EXPANSION JOINT—FIBER

Sizes (Inch Thick):	1/4	3/8	1/2	3/4	1
Weight/100 Sq. Ft. (Lbs.):	70	95	120	190	260
Sheet Sizes: 3 × 5 – 10 feet (Cut to all sizes on order)					

EXPANSION JOINT—SPACING IN CONCRETE

When expansion joints are placed at approximate intervals of	10 to 20 (ft.)	20 to 30	30 to 50	50 to 70	70 to 100
Thickness of joint filler should be	1/4 (in.)	3/8	1/2	3/4	1

EXPANSION JOINT—STRIP SIZES
Weights/100 Cubic Linear Foot

Strip Widths (In.)	Thicknesses				
	1/4 In.	3/8 In.	1/2 In.	3/4 In.	1 In.
ASPHALT TYPE					
2	24	36	47	71	95
2 1/2	30	45	59	89	119
3	35	54	71	107	143
3 1/2	41	63	83	124	156
4	47	72	95	142	190
4 1/2	53	81	106	160	210
5	59	90	118	178	238
5 1/2	65	99	130	195	257
6	71	107	142	213	285
6 1/2	77	116	154	231	309
7	83	125	166	249	327
7 1/2	89	134	177	266	356
8	95	143	189	284	374
8 1/2	101	152	201	302	397
9	106	161	213	320	428
9 1/2	112	170	225	337	444
10	118	179	237	355	475
10 1/2	124	188	248	373	491
11	130	197	260	391	514
11 1/2	136	206	272	408	538
FIBER TYPE					
2	12	18	26	35	52
2 1/2	16	23	32	44	64
3	19	28	39	53	77
3 1/2	22	32	46	62	90
4	25	37	52	70	103
4 1/2	28	41	59	79	116
5	31	46	65	88	129
5 1/2	34	51	72	97	142
6	37	55	78	106	155
6 1/2	40	60	85	114	168
7	44	64	91	123	180
7 1/2	47	69	98	132	193
8	50	74	104	141	206
8 1/2	53	78	111	150	219
9	56	83	117	158	232
9 1/2	59	87	124	167	245
10	62	92	130	176	258
10 1/2	65	97	137	185	271
11	68	101	143	194	284
11 1/2	72	106	150	202	297
12	75	110	156	211	309

F

FASTENERS—CORRUGATED
(SIZES, PACKAGES)

Length (In.)	Corrugations	No./Box
$1/4$	3-4-5	100-500
$3/8$	3-4-5	100-500
$1/2$	3-4-5	100-500
$5/8$	3-4-5	100-500
$3/4$	3-4-5	100-500

FENCE PICKETS

Size	Lengths				
$1'' \times 3''$	3'	3'–6''	4'	5'	6'
Wt./100 pcs.	135	155	175	220	265

FENCE POSTS—SIZES, WEIGHTS

Punched Channel Steel
(1 1/4 × 5/8 In., 14 Ga.)

Lengths (Ft.)	Weight (Lbs.)
3	2
4	2 1/2
5	3
5 1/2	3 1/2
6	4

T Steel Studded
(1 3/8 × 1 3/8 In., 7/64 In. Thick)

Fence Ht. (In.)	Weight (Lbs.)
36	7 1/2
42	8
48	8 1/2
60	10
72	11 1/2

Punched Channel Steel
(2 3/16 × 7/8 In., 1/8 In. Thick)

Weight (Lbs.)
6 1/2
7
8
9
10

Wood Tapered
3 In. Top Dia.—2 3/4 to 3 3/4In.
4 In. Top Dia.—3 1/4 to 4 3/4 In.

Peeled Cedar

Size (In.)	Length (Ft.)	Wt. (Lbs.)
3	7	18
4	7	20

Peeled Cypress

Size (In.)	Length (Ft.)	Wt. (Lbs.)
3	6	18
4	6	22
3	7	22
4	7	25

FINISHES—EFFECTIVENESS

Wood was initially conditioned to 80 degrees Fahrenheit and 65 percent relative humidity and then exposed for 2 weeks to 80 degrees Fahrenheit and 97 percent relative humidity.

Coatings	Effectiveness
INTERIOR FINISHES	**PCT.**
Uncoated wood.	0
3 coats of phenolic varnish.	73
2 coats of phenolic varnish.	49
1 coat of phenolic varnish (sealer).	5
3 coats of shellac	87
3 coats of cellulose lacquer	73
3 coats of lacquer enamel	76
3 coats of furniture wax	8
3 coats of linseed oil.	21
2 coats of linseed oil.	5
1 coat of linseed oil (sealer)	1
2 coats of latex paint.	0
2 coats of semigloss enamel	52
2 coats of floor seal	0
2 coats of floor seal plus wax.	10
EXTERIOR FINISHES	
1 coat water-repellent preservative[1]	0
1 coat of FPL natural finish (penetrating stain)	0
1 coat house paint primer.	20
1 coat of house primer plus 2 coats of latex paint	22
1 coat of house primer, plus 1 coat of TZ[2] linseed oil paint, 30 percent PVC[3]	60
1 coat of house primer plus 1 coat of TL[2] linseed oil paint, 30 percent PVC.	65
1 coat of T-alkyd-oil, 30 percent PVC.	45
1 coat of T-alkyd-oil, 40 percent PVC.	3
1 coat of T-alkyd-oil, 50 percent PVC.	0
2 coats of exterior latex paint	3
1 coat aluminum powder in long oil phenolic varnish	39
2 coats aluminum powder in long oil phenolic varnish	88
3 coats aluminum powder in long oil phenolic varnish	95

[1] The same product measured by immersing in water for 30 min. would have a water-repellency effectiveness of over 60 pct.

[2] The letters *T, L*, and *Z* denote paint's pigment with titanium dioxide, basic carbonated white lead, and zinc oxide, respectively.

[3] PVC denotes pigment volume concentration which is the volume percent of pigment in the nonvolatile portion of the paint.

FILM THICKNESS
Coverage Liquid, Plastic Materials

Thickness of Film (In.)	Coverage/ Gal. (Sq. Ft.)
$1/300$	408
$1/256$	409
$1/128$	205
$1/64$	103
$1/32$	51
$1/16$	26
$1/8$	13
$1/4$	6.4
$1/2$	3.2

FIRE CLAY

- Packaged: 100-pound sacks
- 500 pounds required/1000 standard bricks
- Add 10 – 25% Portland or mortar cement

FIREPLACE—ASH DUMPS

Size (In.)	Wt. (Lbs.)
$4\ 1/2 \times 8$	$2\ 1/2$
5×8	$2\ 1/2$
7×10	12

Various types available

FIREPLACE—ASH PIT DOORS

Size (In.)	Wt. (Lbs.)
8×8	7
8×12	10
12×12	13
18×24	50
24×30	70

Various types available

FIREPLACE—BASKET GRATES

Width (In.)		Depth (In.)	For Opngs (In.)	Wt. (Lbs.)
Front	Back			
24	12	15	24 to 30	43
28	16	15	30 to 35	45
30	18	15	32 to 37	47
34	22	15	37 to 42	55
40	28	15	42 to 58	71
57	45	15	58 to 72	115

Andiron clearance: 4 1/2", Heights: 12"

FIREPLACE—DAMPERS

Fireplace Opening Width (In.)	Throat Size			Weight (Lbs.)
	Bottom Width	Top Width	Opng. Width	
24	24	17 $5/16$	4 $1/4$	27
30	30	23 $5/16$	4 $1/4$	31
33	33	26 $5/16$	4 $1/4$	35
36	36	29 $5/16$	4 $1/4$	37
42	42	35 $5/16$	4 $1/4$	43
48	48	41 $5/16$	4 $1/4$	48
54	54	42 $1/2$	7	90
60	60	49 $1/2$	7	100
72	72	60 $1/2$	7	130
84	84	73 $1/2$	7	155
96	96	85 $3/4$	7	175

FIREPLACES—CIRCULATING

FOR CHIMNEYS 8 TO 20 FEET HIGH

FOR CHIMNEYS 20 TO 40 FEET HIGH

Opening Size (In.)	Wt. (Lbs.)	Standard Flue Size (In.)	Modular Flue Size (In.)	Round Flue Size (In.)	Standard Flue Size (In.)	Modular Flue Size (In.)	Round Flue Size (In.)
28	155	8 1/2 × 13	12 × 12	10	8 1/2 × 13	12 × 12	10
32	180	13 × 13	12 × 16	12	8 1/2 × 13	12 × 12	10
34	200	13 × 13	12 × 16	12	8 1/2 × 13	12 × 12	10
36	225	13 × 13	12 × 16	12	13 × 13	12 × 16	12
40	280	13 × 18	16 × 16	15	13 × 13	12 × 16	12
46	325	13 × 18	16 × 16	15	13 × 13	12 × 16	12
48	365	13 × 18	16 × 20	15	13 × 18	16 × 16	15
52	500	13 × 18	16 × 20	15	13 × 18	16 × 16	15
60	610	18 × 18	20 × 20	18	13 × 18	16 × 20	15

Fireplaces—Flue, Dampers, Angles Required: Fire Ratings—Ceilings; Fire Ratings—Frame Partitions—see page 67.

FIREPLACES—RECOMMENDED SIZES

Room Size (Ft.)	Location of Fireplace			
	On Short Side		On Long Side	
	Width Opng. (In.)	Height Opng. (In.)	Width Opng. (In.)	Height Opng. (In.)
10 × 14	28	27	28	27
12 × 16 (	28	27	32	27
(	32	27	36	27
12 × 20 (	32	27	36	27
(	36	27	40	27
12 × 24 (	32	27	36	27
(	36	27	40	29
14 × 28 (	36	27	40	29
(	40	29	48	29
16 × 30 (	36	27	40	29
(	40	29	48	29
20 × 30 (	40	29	48	29
(	48	29	60	32

FIREPLACES—FLUE, DAMPERS, ANGLES REQUIRED

Front Opening (In.)	Modular (In.)	Old Size (In.)	Round Flue (In.)	Damper Size (In.)	Angle for Fireplace (In.)	Angle for Chimney (In.)
24	8×12	8 1/2×8 1/2	8	24	3×3×3/16 − 36	3×3×3/16 − 36
26	8×12	8 1/2×8 1/2	8	30	3×3×3/16 − 36	3×3×3/16 − 36
28	8×12	8 1/2×13	10	30	3×3×3/16 − 36	3×3×3/16 − 36
30	12×12	8 1/2×13	10	30	3×3×3/16 − 36	3×3×3/16 − 36
32	12×12	8 1/2×13	10	33	3×3×3/16 − 42	3×3×3/16 − 42
36	12×12	13×13	12	36	3×3×3/16 − 42	3×3×3/16 − 42
40	12×16	13×13	12	42	3×3×3/16 − 48	3×3×3/16 − 48
42	16×16	13×13	12	42	3 1/2×3×1/4 − 48	3×3×3/16 − 48
48	16×16	13×13	15	48	3 1/2×3×1/4 − 54	3 1/2×3×1/4 − 54
54	16×16	13×18	15	54	3 1/2×3×1/4 − 60	3 1/2×3×1/4 − 60
60	16×20	13×18	15	60	3 1/2×3×1/4 − 66	3 1/2×3×1/4 − 72
72	16×20	13×18	18	72	3 1/2×3×1/4 − 66	3 1/2×3×1/4 − 72

FIRE RATINGS—CEILINGS

Const.	Base	Plaster & Mix	Plaster Thickness	Rating
Wood frame	Metal lath	Gyp. 1:2, 1:3	3/4"	45 min.
Wood frame	Metal lath	Gyp. 1:2, 1:2	3/4"	1 hour
Wood frame	Metal lath	Gyp. wood fiber	3/4"	1 hour
Wood frame	Perf. gyp. lath	Gyp. 1:2	1/2"	45 min.

Check local codes and current ratings

FIRE RATINGS—FRAME PARTITIONS

Type	Base	Plaster Mix, Coats	Plaster Thickness	Rating
Wood frame	3/8" Plain gypsum lath	Gypsum 1:2, 1:2	1/2"	45 min.
Wood frame	3/8" Plain gypsum lath	Gypsum wood fiber	1/2"	1 hour
Wood frame	3/8 Perf. gypsum lath	Gypsum 1:2, 1:2	1/2"	1 hour
Wood frame	Metal lath	Gypsum 1:3, 1:2	3/4"	45 min.
Wood frame	Metal lath	Gypsum 1:2, 1:2	3/4"	1 hour
Wood frame	Metal lath	Gypsum wood fiber	3/4"	1 1/2 hr.

Check local codes and current ratings

FIRE RATINGS—GYPSUM LATH (PLASTER)

Gypsum Lath Base (In.)	Mix Plaster: Aggregate	Plaster Thickness (In.)	Resistance Rating	
			Bearing	Non-Bearing
PARTITIONS				
2 × 4 Wood Frame				
3/8 plain	1:2 sand	1/2	45 min.	
3/8 plain	Wood fiber	1/2	1 hr.	
3/8 perf.	1:2 sand	1/2	1 hr.	
3/8 perf.	100:2 1/2 vermiculite or perlite	1/2	1 hr.	1 1/2 hrs.
SOLID STUDLESS				
1/2 plain	1:1, 1:2 sand	2 (includes lath)		1 hr.
1/2 plain	100:2, 100:3 verm./perl.	2 (inc. lath)		1 1/2 hrs.
1/2 plain	100:2, 100:3 vermiculite	2 1/2 (inc. lath)		2 hrs.
FLOOR & CEILING CONSTRUCTION				
(Wood Joists, 16″ O.C.)				
3/8 perf.	100:2 1/2 perl. or verm.	1/2	(1)	1 1/2 hrs.
3/8 perf.	100:2 1/2 perl. or verm.	1/2		1 hr.
3/8 perf.	1:2 sand	1/2	(2)	1 hr.
3/8 perf.	1:2 sand	1/2		45 min.
(Steel Joists, 24″ O.C.)				
3/8 perf.	100:2 1/2 perlite	5/8		1 hr.
3/8 perf.	100:2 1/2 perlite	5/8	(3)	3 hrs.
3/8 perf.	100:2 1/2 perlite	1/2	(1)	3 hrs.
3/8 perf.	100:2 1/3 perlite	1	(1)	4 hrs.
METAL COLUMNS				
3/8 perf.	1:2 1/2 sand	1/2		1 hr.
3/8 perf.	1:2 1/2 sand	5/8		1 1/2 hrs.
	100:2 perlite or vermiculite	1 3/4		2 hrs.
3/8 perf.	100:2, 100:3 perlite	1 3/4		3 hrs.
1/2 full length 2-layers	100:2, 100:3 perlite	1 1/2	(1)	4 hrs.

FIRE RATINGS—MASONRY (CONCRETE)

Type of Construction	Minimum Thickness (In.) for Ratings Shown			
	4 Hour	3 Hour	2 Hour	1 Hour
Expanded slag or pumice	4.7	4.0	3.2	2.1
Expanded clay or shale	5.7	4.8	3.8	2.6
Limestone, cinders or unexpanded slag.............	5.9	5.0	4.0	2.7
Gravel.......................	6.2	5.3	4.2	2.8
Silica gravel	6.7	5.7	4.5	3.0

Check local codes

FIRE RATINGS—MASONRY (CONCRETE)

Wall Thickness (In.)	Unit Type	% Solid (Min.)	Materials			
			Exp. Slag or Pumice		Limestone, Slag, Cinders Exp. Shale	
			No Plaster	Plaster 2 Sides	No Plaster	Plaster 2 Sides
3	Partition	72	1 hr.	2 hr.	3/4 hr.	1 3/4 hr.
4	Partition	61	1 hr.	2 1/2 hr.	1 hr.	2 hr.
4	Partition	72	2	2 1/2	1 1/4	2
6	Partition	49	1 1/2	2 1/2	1 1/4	2
6	Partition	60	2	3 1/2	1 1/2	2 1/2
6	Partition	75	3	5	2 1/2	4
6	Load Bearing	70	3	5	—	—
6	Load Bearing	75	3 1/2	5	—	—
8	Load Bearing	55	3	5	2	3 1/2
8	Load Bearing	62	4	6	2 1/2	4
10	Load Bearing	60	5	7	4	6
12	Load Bearing	55	—	—	4	6

Check local codes and current ratings

FIRE RATINGS—METAL LATH

CEILINGS

Const.	Plaster	Thick (In.)	Rating
Wood Frame	Gypsum-Sand, 1:2, 1:3	3/4	1 Hour
Steel Joists	Gypsum-Sand, 1:2, 1:3	3/4	2 Hour
Steel Joists	Gypsum-Vermiculite, 100:2, 100:3	3/4	3 Hour
Steel Joists	Gypsum-Wood Fiber	1	3 Hour
Steel Joists	Gypsum-Vermiculite, 100:2, 100:3	1	4 Hour
Cellular Steel Floor	Gypsum-Wood Fiber	1	4 Hour
Cellular Steel Floor	Gypsum-Vermiculite, 100:2, 100:3	1	4 Hour
Cellular Steel Floor	Gypsum-Perlite, 100:3, 100:3	1	4 Hour
Suspended Channel	Gypsum-Vermiculite, 100:2, 100:3	1	4 Hour

COLUMNS

Const.	Plaster	Thick (In.)	Rating
Steel Section	Gypsum-Sand, 1:3, 1:3	3/4	1 Hour
Steel Section	Gypsum-Vermiculite, 100:2, 100:3	1	3 Hour
Steel Section	Gypsum-Vermiculite, 100:2, 100:3	1 1/2	4 Hour
Steel Section	Gypsum-Perlite, 100:2, 100:3	1	3 Hour
Steel Section	Gypsum-Perlite, 100:2, 100:3	1 1/2	4 Hour

PARTITIONS

Const.	Plaster	Thick (In.)	Rating
Wood	Gypsum-Sand, 1:2, 1:3	3/4	45 Minutes
Wood	Gypsum-Sand, 1:2, 1:2	3/4	1 Hour
Wood	Gypsum-Sand, 1:2, 1:3	7/8	1 Hour
Wood	Gypsum-Wood Fiber	3/4	1 1/2 Hour
Wood	Portland Cement-Sand, 1:2, 1:3	3/4	30 Minutes
Wood	Portland Cement-Sand, 1:2, 1:3	7/8	45 Minutes
Wood	Portland Cement-Lime-Sand, 1:2:8, 1:2:10	3/4	30 Minutes
Wood	Gypsum-Vermiculite 100:2 1/2, 100:3 1/2	3/4	1 Hour
Solid Plaster	Gypsum-Perlite 100:1 1/2, 100:1 1/2	1/2	1 Hour
Solid Plaster	Gypsum-Sand, 1:2, 1:2	2	1 Hour
Solid Plaster	Gypsum-Wood Fiber	2 1/4	2 Hour
Studs	Gypsum-Sand, 1:2, 1:3	3/4	45 Minutes
Studs	Gypsum-Sand, 1:2, 1:2	3/4	1 Hour
Studs	Gypsum-Sand, 1:2, 1:3	7/8	1 Hour
Studs	Gypsum-Wood Fiber	7/8	2 Hour
Studs	Gypsum-Perlite 100:2, 100:3	1	2 Hour

Check local codes and current ratings

FIRE RATINGS—PLASTER

Construction	Type of Base	Gypsum Plaster to Sand, by Weight	Thickness (In.)	Fire Rating
PARTITIONS				
Wood Frame	3/8" Gypsum Lath, Plain	1:2, 1:2	1/2	45 minutes
Wood Frame	3/8" Gypsum Lath, Plain	Gypsum Wood Fiber Plaster (Neat)	1/2	1 hour
Wood Frame	3/8" Gypsum Lath, Perf.	1:2, 1:2	1/2	1 hour
Wood Frame	Metal Lath	1:2, 1:3	3/4	45 minutes
Wood Frame	Metal Lath	1:2, 1:2	3/4	1 hour
Wood Frame	Metal Lath	Gypsum Wood Fiber Plaster (Neat)	3/4	1 1/2 hours
Solid	Metal Lath	1:2, 1:3	2	45 minutes
Solid	Metal Lath	1:2, 1:2	2	1 hour
Solid	Metal Lath	Gypsum Wood Fiber Plasters (Neat)	2 1/4	2 hours
Solid (Studless)	Metal Lath	1:2, 1:2	2	1 hour
Solid (Studless)	Gypsum Lath	1:1, 1:2	2	1 hour
Wood Frame	3/8" Gypsum Lath, Perf.	Gypsum Plaster, 100 lbs. To: 2 1/2 cu. ft. Perlite or Vermiculite	1/2	1 1/4 hours
Wood Frame	Metal Lath	2 1/2:3 1/2 cu. ft. Vermiculite	3/4	1 hour
Solid	Metal Lath	2 1/2 cu. ft. Perlite	1 1/2	1 hour
Solid	Metal Lath	2:3 cu. ft. Perlite or Vermiculite	2 1/2	2 hours
Solid (Studless)	Gypsum Lath	2:3 cu. ft. Perlite or Vermiculite	3/4	1 1/2 hours
CEILINGS				
Wood Frame	3/8" Gypsum Lath, Perf.	1:2	1/2	45 minutes
Wood Frame	3/8" Gypsum Lath, Perf.	1:2	1/2	1 hour
Wood Frame	Metal Lath	1:2, 1:3	3/4	45 minutes
Wood Frame	Metal Lath	1:2, 1:3	3/4	1 hour
Wood Frame	Metal Lath	Gypsum Wood Fiber (Neat)	3/4	1 hour
Wood Frame	3/8" Gypsum Lath, Perf.	Gypsum Plaster, 100 lbs. To: 2 1/2 cu. ft. Perlite	1/2	1 hour

Check local codes and current ratings

FIRE RATINGS—STEEL & PLASTER PARTITIONS

Studs	Total Thickness (In.)	Thickness and Type of Plaster on Metal Lath (In.)	Fire Resistance Rating
3/4" Channels	2 1/2	2 1/2 neat wood-fibered gypsum	2 1/2 hours
	2 1/2	2 1/2 gypsum-perlite or vermiculite	2 hours
	2	2 neat wood-fibered gypsum	2 hours
	2	2 gypsum-sanded 1:2, 1:2	1 hour
	1 1/2	1 1/2 gypsum-perlite	1 hour
Studless	2	2 gypsum-sanded 1:2, 1:2	1 hour

Check local codes and current ratings

FIRE RATINGS—WOOD WALLS & PARTITIONS
Based on ASTM E119, except where noted

BEARING WALLS—2 x 4 STUDS SPACED 16" O.C.

Faces	Insulation	Rating, Min.	Notes
1. 1/4" plywood, 3/8" plywood	3 1/2" Fiberglass	15	600 plf load on 8 ft. high wall
2. 3/8" gypsum, 5/8" T1–11 plywood	3 1/2" Fiberglass	30	1,250 plf load on 8 ft. high wall
3. 3/8" plywood, both faces	Full thick, mineral wool	30	(40 minutes, nonbearing rating)
4. 5/8" Type X gypsum, 5/8" plywood	3 1/2" Mineral wool	45	FRT studs and plywood
5. Same as 4	3 1/2" Mineral wool	60	Same; but plywood nailed to studs through 4" gypsum strips

6. 3/8" gypsum both faces	None	30	
7. 1/2" gypsum, both faces	None	45	
8. 5/8" Type X gypsum both faces	None	60	
9. 2 layers 5/8" Type X gypsum, both faces	None	120	
NONBEARING WALLS—1 × 3 STUDS SPACED 16" O.C.			
10. 1/4" plywood, both faces	None	10	
11. 3/8" plywood, both faces	None	20	Values obtained on small size walls
12. 1/2" plywood, both faces	None	25	
13. 5/8" plywood, both faces	None	35	
14. 1/4" plywood, 3/8" plywood	Mineral wool 2 lbs./sq. ft.	50	8 ft. high wall

Check local codes and current ratings

FLAG POLES—STEEL (SIZES, WEIGHTS)

Exposed Height (Ft.)	Setting Depth (Ft.)	Total Length (Ft.)	Outside Diameter			Weight (Lbs.)	Tapered Wall (In.)
			Butt	Top	Ball		
20	3	23	5	3 1/4	5	325	.250
25	3 1/2	28 1/2	5 9/16	3 1/4	6	430	.250
30	3 1/2	33 1/2	6	3 1/4	6	535	.250
35	4	39	6 5/8	3 1/4	6	700	.250
40	4	44	7 5/8	3 1/4	8	870	.250
50	5	55	8 5/8	3 1/4	8	1235	.250
59	6	65	10 3/4	3 1/4	10	1550	.250
70	7	77	11 3/4	3 1/4	10	2400	.250
75	7 1/2	82 1/2	12 3/4	4	12	2850	.250
80	8	88	14	4	12	3475	.250
90	9	99	15	4	14	4225	.250
100	10	110	16	4	14	5000	.250

FLAGS—RECOMMENDED SIZES

Height of Pole (Ft.)	Flag Size (Ft.) If Pole Top Is 2 3/8 In. or Smaller	Flag Size (Ft.) If Pole Top Is 2 7/8 In. or Larger
20	3 × 5	4 × 6
25	3 × 5	4 × 6
30	4 × 6	5 × 8
40	5 × 8	6 × 9
50	6 × 9	6 × 10
60	6 × 10	8 × 12
70	8 × 12	10 × 15
75	8 × 12	10 × 15
80	8 × 12	10 × 15
90	10 × 15	10 × 19
100	10 × 15	12 × 20

FLASHING—ALUMINUM

ROLLS

Width (In.)	Thickness (In.)	Length Per Roll (Ft.)	Weight Per Roll (Lbs.)
14	.019	50	16
20	.019	50	23
20	.024	50	32
28	.019	50	29

ROLLS

Length (Ft.)	Width (In.)	Thickness (In.)	Weight (Lbs.)	Packing Carton
50	14	.019	16	4
50	20	.019	23	4
50	20	.024	32	4
50	28	.019	29	2

SHEETS

Width (In.)	Length (Ft.)	Thickness
18	4	26 ga.
28	6-8-10-12	26 ga.

SHEETS

Length (In.)	Width (In.)	Thickness (In.)	Weight (Lbs.)	Packing Carton
48	18	.024	4	10

Specific products may vary

FLOORING—HARDWOOD
Number of Board Feet of Standard Sizes
Flooring to Cover Various Square Feet Amounts

Square Feet Floor	25/32 × 3 1/4	25/32 × 2 1/4	25/32 × 1 1/2	1/2 × 2 1/2	1/2 × 2	3/8 × 2	3/8 × 1 1/2
10	12	13	15	12	12	12	13
20	25	27	30	24	25	25	27
30	38	40	45	36	38	38	40
40	50	53	60	48	50	50	53
50	63	67	75	60	63	63	67
60	75	80	90	72	75	75	80
70	88	93	105	84	88	88	93
80	100	107	120	96	100	100	107
90	113	120	135	108	113	113	120
100	125	133	150	120	125	125	133
200	250	267	300	240	250	250	267
300	375	400	450	360	375	375	400
400	500	533	600	480	500	500	533
500	625	667	750	600	625	625	667
600	750	800	900	720	750	750	800
700	875	933	1050	840	875	875	933
800	1000	1067	1200	960	1000	1000	1067
900	1125	1200	1350	1080	1125	1125	1200
1000	1250	1333	1500	1200	1250	1250	1333
Counted as	1 × 4	1 × 3	1 × 2 1/4	1 × 3	1 × 2 1/2	1 × 2 1/2	1 × 2
Weight	2250	2000	2000	1500	1300	1000	1000
Size of Nails used	8d	8d	8d	6d	6d	4d	4d
Pieces per Bundle	6	12	12	16	18	24	24
To Obtain Bd. Ft. from Bdle. Ft. Multiply by	2	3	2 1/4	4	3 3/4	5	4
To Obtain Bdle. Ft. from Bd. Ft. Multiply by	1/2	1/3	4/9	1/4	4/15	1/5	1/4

FLOORING—HARDWOOD
Estimating Required Amounts of Hardwood Flooring

Calculate the number of sq. ft. to be covered by multiplying width and length of room. To convert this basic sq. ft. amount to the desired end, use the figures in the table opposite the size flooring being considered and in the column headed by the letter:

A—To obtain sq. ft. of flooring required to cover, add percentage shown in this column to room area (includes 3% waste allowance).

B—To obtain board ft. required from sq. ft. of room area, multiply latter by factor in this column (includes 3% waste allowance).

C—To obtain price in cents per sq. ft. multiply price in dollars per M by the factor in this column.

D—To obtain nails required, in pounds, multiply room area by factor in this column.

E—Distance in nailing for various sizes of flooring upon which the factor in column D is based.

Size	A	B	C	D	E
25/32 × 2 1/4	36%	1.36	.136	.045	14 – 16 in.
25/32 × 1 1/2	53%	1.53	.153	.050	14 – 16 in.
1/2 × 2	28%	1.28	.128	.016	10 – 12 in.
1/2 × 1 1/2	36%	1.36	.136	.018	10 – 12 in.
3/8 × 2	28%	1.28	.128	.016	8 – 10 in.
3/8 × 1 1/2	36%	1.36	.136	.018	8 – 10 in.

FLOORING—HARDWOOD BLOCK

Strip Width	1/2 Inch Thick	25/32 Inch Thick	33/32 Inch Thick
1 1/2"	7 1/2" × 7 1/2" 9 × 9 6 × 12	7 1/2" × 7 1/2" 9 × 9 6 × 12	7 1/2" × 7 1/2" 9 × 9 6 × 12
2"	8 × 8 10 × 10 6 × 12	8 × 8 10 × 10 6 × 12	8 × 8 10 × 10 6 × 12
2 1/4"	- - - - - - - - - - - -	6 3/4 × 6 3/4 9 × 9 11 1/4 × 11 1/4 6 3/4 × 13 1/2	6 3/4 × 6 3/4 9 × 9 11 1/4 × 11 1/4 6 3/4 × 13 1/2

FLOORING—HARDWOOD GRADES

UNFINISHED OAK FLOORING (Red & White Separated)		UNFINISHED HARD MAPLE (Beech & Birch*)		

CLEAR PLAIN or CLEAR QUARTERED*
Best appearance.
Best grade, most uniform color, limited small character marks.
Bundles 1 1/4 ft. and up. Average length 3 3/4 ft.
	SELECT AND BETTER	**FIRST GRADE**
		Best appearance.
		Natural color variation, limited character marks, unlimited sap.
		Bundles 2 ft. & up.
		2 & 3 ft. bundles up to 33% footage.
		  FIRST GRADE WHITE HARD MAPLE
		  (Special Order)
		  FIRST GRADE RED BEECH & BIRCH
		  (Special Order)
			SECOND AND BETTER	**THIRD AND BETTER**

SELECT PLAIN or SELECT QUARTERED* Excellent apperance. Limited character marks, unlimited sound sap. Bundles 1 1/4 ft. and up. Average length 3 1/4 ft. **SELECT & BETTER*** A combination of Clear and Select grades.		**SECOND GRADE** Variegated appearance. Varying sound wood characteristics of species. Bundles 2 ft. & up. 2 & 3 ft. bundles up to 45% footage. **SECOND & BETTER GRADE** A combination of FIRST and SECOND GRADES. Bundles 2 ft. & up. 2 & 3 ft. bundles up to 40% footage.		
NO. 1 COMMON Variegated appearance. Light and dark colors; knots, flags, worm holes, and other character marks allowed to provide a variegated appearance after imperfections are filled and finished. Bundles 1 1/4 ft. and up. Average length 2 3/4 ft.		**THIRD GRADE** Rustic appearance. All wood characteristics of species. Serviceable economical floor after filling. Bundles 1 1/4 ft. and up. 1 1/4 ft. to 3 ft. bundles as produced up to 65% footage. **THIRD & BETTER GRADE** A combination of FIRST, SECOND and THIRD GRADES. Bundles 1 1/4 ft. and up. 1 1/4 ft. to 3 ft. bundles as produced up to 50% footage.		
NO. 2 COMMON Rustic appearance. All wood characteristics of species. A serviceable economical floor after knot holes, worm holes, checks, and other imperfections are filled and finished. Bundles 1 1/4 ft. and up. Average length 2 1/4 ft. Red and White may be mixed.				
1 1/4' SHORTS Pieces 9 to 18 inches. Bundles average nominal 1 1/4 ft. **NO. 1 COMMON & BETTER SHORTS** A combination grade of CLEAR, SELECT, & NO. 1 COMMON **NO. 2 COMMON SHORTS** Same as No. 2 Common.		A brief grade description, for comparison only. NOFMA flooring is bundled by averaging the lengths. A bundle may include pieces from 6 inches under to 6 inches over the nominal length of the bundle. No piece shorter than 9 inches admitted. The percentages under 4 ft. referred to apply on total footage in any one shipment of the item. 3/4 inch added to face length when measuring length of each piece. NESTED FLOORING is random length flooring bundled end to end continuously in 8 ft. long [nominal] bundles. OAK Regular grade requirements apply. BEECH. BIRCH. HARD MAPLE & PECAN 9 – 18 inch pieces will be admitted in 3/4 x 2 1/4" as follows: First Grade = 4 pcs. Second Grade = 8 pcs. Third Grade = as develops. Average Lengths; First Grade = 42 Inches. Second Grade = 33 Inches. Third Grade = 30 Inches.		

(NOTE: Flooring specified "QUARTERED" shall contain Quartered only.

Flooring specified "PLAIN" *may* contain both Plain and Quartered.)

*Check for availability.

UNFINISHED PECAN FLOORING*	PREFINISHED OAK FLOORING (Red & White separated, graded after finishing)		
FIRST GRADE Excellent appearance. Natural color variation, limited character marks, unlimited sap. Bundles 2 ft. & up. 2 & 3 ft. bundles up to 25% footage. FIRST GRADE RED (Special Order) FIRST GRADE WHITE (Special Order)	**PRIME GRADE** (Special Order) Excellent appearance. Natural color variation, limited character marks, unlimited sap. Bundles 1 1/4 ft. & up. Average length 3 1/2 ft.		
SECOND GRADE Variegated appearance. Varying sound wood characteristics of species. Bundles 1 1/4 ft. and up. 1 1/4 ft. to 3 ft. bundles as produced up to 40% footage. SECOND & BETTER GRADE A combination of FIRST and SECOND GRADES.	**STANDARD GRADE** Variegated appearance. Varying sound wood characteristics of species. A sound floor. Bundles 1 1/4 ft. & up. Average length 2 3/4 ft. STANDARD & BETTER GRADE Combination of STANDARD and PRIME. Bundles 1 1/4 ft. & up. Average length 3 ft.	STANDARD AND BETTER	TAVERN AND BETTER
THIRD GRADE Rustic appearance. All wood characteristics of species. A serviceable, economical floor after filling. Bundles 1 1/4 ft. and up. 1 1/4 ft. to 3 ft. bundles as produced up to 60% footage. THIRD & BETTER GRADE A combination of FIRST, SECOND and THIRD GRADES.	**TAVERN GRADE** Rustic appearance. All wood characteristics of species. A serviceable, economical floor. Bundles 1 1/4 ft. & up. Average length 2 1/4 ft. TAVERN & BETTER GRADE (Special Order) Combination of PRIME, STANDARD and TAVERN. All wood characteristics of species. Bundles 1 1/4 ft. & up. Average length 3 ft.		
	PREFINISHED BEECH & PECAN FLOORING TAVERN & BETTER GRADE (Special Order) Combination of PRIME, STANDARD and TAVERN. All wood characteristics of species. Bundles 1 1/4 ft. & up. Average length 3 ft.		

FLOORING—HARDWOOD STRIP

Computing Board Foot Content of Hardwood Flooring Bundles
(For example: One 6-Foot Bundle of $3/4 \times 2 \ 1/4$ Inch Flooring Pack-
aged 12 Pieces per Bundle Yields 18 Board Feet

Size (In.)	Number of Pieces in Bundle	Multiply Bundle Length by
$3/4 \times 1 \ 1/2$	12	$2 \ 1/4$
	8	$1 \ 1/2$
	16	3
$3/4 \times 2$	12	$2 \ 3/4$
	8	$1 \ 5/6$
$3/4 \times 2 \ 1/4$	12	3
	8	2
$3/4 \times 3 \ 1/4$	8	$2 \ 2/3$
$3/8 \times 1 \ 1/2$	24	4
$3/8 \times 2$	24	5
$1/2 \times 1 \ 1/2$	18	3
$1/2 \times 2$	18	$3 \ 3/4$
$5/16 \times 1 \ 1/2$	24	3
$5/16 \times 2$	24	4

FLOORING—HARDWOOD STRIP

Estimating the Amount of Hardwood Strip Flooring Required

An allowance for side-matching (plus 5% for end-matching and normal waste) is incorporated in these percentages.

Take the square footage and ADD the percentage below opposite the size strip flooring to be used.

Strip (When Using)	Add
$3/4 \times 1 \ 1/2''$	55%
$3/4 \times 2$	$42 \ 1/2$
$3/4 \times 2 \ 1/4$	$38 \ 1/3$
$3/4 \times 3 \ 1/4$	29
$3/8 \times 1 \ 1/2$	$38 \ 1/3$
$3/8 \times 2$	30
$1/2 \times 2 \ 1/2$	$38 \ 1/3$
$1/2 \times 2$	30

Above percentages are for laying flooring straight across the room. Additional flooring should be estimated for diagonal applications and bay windows or other projections.

FLOORING—HARDWOOD STRIP

Converting Square Feet of Floor Space to Board Feet of Strip Flooring Required.

Floor Space	Board Feet Required (5% Cutting Waste Included)				
Square Feet	3/4 × 2 1/4″	3/4 × 1 1/2″	3/4 × 3 1/4″	1/2 × 2″	3/8 × 1 1/2″
5	7	8	6	7	7
10	14	16	13	13	14
20	28	31	26	26	28
30	42	47	39	39	42
40	55	62	52	52	55
50	69	78	65	65	69
60	83	93	77	78	83
70	97	109	90	91	97
80	111	124	103	104	111
90	125	140	116	117	125
100	138	155	129	130	138
200	277	310	258	260	277
300	415	465	387	390	415
400	553	620	516	520	553
500	692	775	645	650	692
600	830	930	774	780	830
700	968	1085	903	910	968
800	1107	1240	1032	1040	1107
900	1245	1395	1161	1170	1245
1000	1383	1550	1290	1300	1383

FLOORING—HARDWOOD STRIP (SIZES)

	Nominal	Actual
	STANDARD SIZES (INCH)	
Tongue & Groove End Matched	3/4 × 2 1/4	3/4 × 2 1/4
	3/4 × 1 1/2	3/4 × 1 1/2
	1/2 × 2	15/32 × 2
	1/2 × 1 1/2	15/32 × 1 1/2
Limited Production	3/4 × 3 1/4	3/4 × 3 1/4
	SPECIAL ORDER SIZES (INCH)	
Tongue & Groove End Matched	3/4 × 2	3/4 × 2
	3/8 × 2	11/32 × 2
	3/8 × 1 1/2	11/32 × 1 1/2
Square Edge	5/16 × 2	5/16 × 2
	5/16 × 1 1/2	5/16 × 1 1/2

FLOORING—PORCH (REINFORCED CONCRETE)

Span (Ft.)	Floor Thickness (In.)	Size of Bars (In.)	Spacing of Bars (In.)
4	4	$1/4$	8
5	4	$1/4$	6
6	5	$3/8$	9
8	6	$3/8$	6
10	6	$1/2$	4

Check local codes

FLOORING—PREFINISHED HARDWOOD
Tongue & Groove, End-Matched

Widths (In.)	Thickness (In.)	Lengths (In.)
3, 5, 7	$3/8$	12 to 60
2	$3/8$	10 to 60
$2^1/4$	$3/4$	12 to 84

Packaged: 15, 20, 25 sq. ft. Cartons
Adhesives required: 1 gal./50 sq. ft.
Check suppliers for plank type, parquet; or other patterns available.

FLOORING—PREFINISHED HARDWOOD BLOCK

- Size—12 × 12 in.
- Thickness—$5/16$ in.
- Availability—plain, self-sticking, various patterns.
- Adhesives—required for plain types, 1 gal./50 sq. ft.

FLOORING—RESILIENT
(TYPICAL SIZES & GAUGES)

Type	Sizes	Gauges (In.)
Sheet, vinyl	3'0", 6'0", 6'6", 9'0", & 12'0' wide	0.065 – 0.220
Tile, cork	6"×6", 9"×9", 6"×12", 12"×12"	1/8, 3/16, 1/4, 5/16, 1/2
Tile, rubber	9"×9", 12"×12", 18"×36", 36"×36"	0.080, 3/32, 1/8. 3/16
Tile, vinyl— (solid, backed)	9"×9", 12"×12"	1/16, 3/32, 1/8 0.045, 0.050, 0.080, 0.095
Tile, vinyl— (composition)	9"×9", 12"×12"	1/16, 3/32, 1/8 0.045, 0.080

Vinyl sheet made with filled or unfilled surface, wear surface thickness ranges from 0.006 to 0.050 in. Special or different sizes of all products may be available.

FLOORING—SOFTWOOD
(SOUTHERN PINE)

Nominal Widths (In.)	Nominal Thicknesses (In.)
3/8	2
1/2	3
5/8	4
3/4	5
1	6
1 1/4	
1 1/2	

Lengths 1 to 8 ft. or longer, end-matched.

FLOORS—CONCRETE JOIST (SIZES, WEIGHTS)

Weights of Concrete Joist Floors with 2 1/2-Inch Slab—Pounds per Square Foot.

Depth of Joist only (In.)	20-In. Pans				30-In. Pans			12-In. Units			16-In. Units		
	Width of Joist												
	4	5	6	7	5	6	7	4	5	6	4	5	6
4	....	....	....	....	....	....	....	55	57	58	54	55	56
6	44	46	48	....	41	43	....	66	68	71	61	62	64
8	49	52	55	....	46	48	....	79	81	84	67	70	72
10	54	58	62	....	50	53	....	88	93	96	74	77	81
12	60	65	69	73	55	58	61	98	103	107	80	85	89
14	....	71	76	81	60	63	67	....	....	....	....	....	....

Check local codes

FLUE LINING—MODULAR (SIZES, WEIGHTS)

Inside Area (Sq. In.)	Dimen-sions (In.)	Outside Dimensions (In.)	Length (In.)	Wall Thickness (In.)	Outside Corner Radius (In.)	Weight/ Ft. (Lbs.)
15	4× 8	3.5× 7.5	24	0.5	1	10
20	4×12	3.5×11.5	24	0.625	1	17
27	4×16	3.5×15.5	24	0.75	1	21
35	8× 8	7.5× 7.5	24	0.625	2	17
57	8×12	7.5×11.5	24	0.75	2	26
74	8×16	7.5×15.5	24	0.875	2	34
87	12×12	11.5×11.5	24	0.875	3	33
120	12×16	11.5×15.5	24	1.	3	42
162	16×16	15.5×15.5	24	1.125	4	64
208	16×20	15.5×19.5	24	1.25	4	102
262	20×20	19.5×19.5	24	1.375	5	115
320	20×24	19.5×23.5	24	1.5	5	122
385	24×24	23.5×23.5	24	1.625	6	

FLUE LINING—ROUND (SIZES, WEIGHTS)

Inside Dia. (In.)	Inside Area (Sq. In.)	Approx. Weight/ Foot (Lbs.)	Length (Ft.)	Thickness (In.)
6	28	11	2	5/8
7	39	16	2	11/16
8	50	17	2	3/4
10	79	27	2	7/8
12	113	37	2	1
15	177	52	2	1 1/8
18	254	67	2	1 1/4
20	314	90	2	1 3/8
22	380	110	2	1 5/8
24	452	120	2	1 5/8
27	572	186	2 1/2, 3	2
30	707	218	2 1/2, 3	2 1/8
33	855	315	2 1/2, 3	2 1/4
36	1017	330	2 1/2, 3	2 1/2

FLUE LINING—STANDARD (SIZES, WEIGHTS)

Outside Dimensions (In.)	Inside Area (Sq. In.)	Weight/ Foot (Lbs.)	Length (Ft.)	Thickness (In.)
4 1/2 × 8 1/2	24	11	2	5/8
4 1/2 × 13	38	18	2	5/8
7 1/2 × 7 1/2	39	14	2	5/8
8 1/2 × 8 1/2	52	18	2	5/8
8 1/2 × 13	80	27	2	3/4
8 1/2 × 18	110	36	2	7/8
13 × 13	127	35	2	7/8
13 × 18	183	45	2	7/8
18 × 18	248	70	2	1 1/8
20 × 20	298	103	2	1 3/8
20 × 24	357	115	2	1 1/2
24 × 24	441	129	2	1 1/2

FLUE RINGS

	CLAY			METAL	
Inside Dia. (In.)	**Lengths (In.)**		**Size (In.)**	**Wt. (Lbs.)**	
5	4 1/2, 6, 9, 12		4	1 1/2	
6	4 1/2, 6, 9, 12		5	1 1/2	
7	4 1/2, 6, 9, 12		6	2	
8	4 1/2, 6, 9, 12		7	2 1/4	
9	4 1/2, 6, 9, 12		8	4	
10	4 1/2, 6, 9, 12		9	5	
12	4 1/2, 6, 9, 12		10	6	
			12	10	

FLUES—RECOMMENDED MINIMUM SIZES

CAPACITY			TYPES FLUES—MIN.				
Warm Air (Sq. In. Leader Pipe)	**Radiation** Steam (Sq. Ft.)	Hot Water (Sq. Ft.)	**Round Lining Inside Dimen. (In.)**	**Rectangular Lining Outside Dimen. (In.)**	**Unlined Flue Inside Dimen. (In.)**	**Effective Flue Area Required (Sq. Ft.)**	**Outside Dimen. of Lined Chimney (In.)**
790	590	973	10	8 1/2 × 13	12 × 12	70	17 × 21
1000	690	1140	10	13 × 13	12 × 12	78.5	20 × 20
	900	1490	12	13 × 13	12 × 16	100	21 × 21
	1100	1820	12	13 × 17 1/2	12 × 16	113	21 × 26
	1700	2800	15	13 × 17 1/2	16 × 20	150	21 × 26
	1940	3200	15	17 1/2 × 17 1/2	16 × 20	177	25 × 25
	2130	3520		17 1/2 × 17 1/2	16 × 20	183	26 × 26
	2480	4090	18	20 × 20	20 × 20	234	23 × 23
	3150	5200	18	20 × 24	20 × 20	254	33 × 33
	4300	7100	20	24 × 24	20 × 24	314	39 × 39
	5000	8250		24 × 24	24 × 24	346	40 × 40

Check local codes

FOOTINGS—COLUMN
Size in Inches

Total Design Load (Lbs.)	Allowable Soil Bearing Capacity, (psf)			
	1500	2000	2500	3000
5,000	22 × 22	19 × 19	17 × 17	16 × 16
10,000	31 × 31	27 × 27	24 × 24	22 × 22
15,000		33 × 33	30 × 30	27 × 27
20,000			34 × 34	31 × 31

FOOTINGS—DIMENSION FOR MASONRY WALLS

- Width—2 × wall thickness
- Thickness—same as wall thickness
  (Deep enough to meet local codes)

FOOTINGS—WALL
Width in Inches

Total Design Load (Lbs./LF Footing)	Allowable Soil Bearing Capacity, (psf)			
	1500	2000	2500	3000
1,000	8	6	4.8	4
1,500	12	9	7.2	6
2,000	16	12	9.6	8
2,500	20	15	12	10

FORMS—CONCRETE (STEEL RIB)

Type	Lengths	Thickness	Gauge	Overall Width (In.)	Coverage (In.)	Corrugations Depth (In.)	Corrugations Width (In.)
Std. Wt.	6'-3"–8'-3" 10'-3"–12'-3"	.0156	28	25 1/4	24	9/16	2 13/32
Hvy. Duty	6' to 18'-6" in 6" increments	.021	26	28 3/4	27	7/8	3 3/8
Ex. Hvy. Duty	6' to 21'-6" in 6" increments	.024 .030 .036	24 22 20	29 1/4	27	1 5/16	4 1/2

FORMS—CONCRETE (SUPPORT STUD SPACING)

Concrete Pressure (Lbs./Square Foot)	Approx. Stud Spacing in Inches with Sheathing Thickness of:			
	1"	1 1/4"	1 1/2"	2"
200	24	32	38	—
400	19	26	31	38
600	16	22	27	34
800	15	20	25	31
1000	14	19	23	28
1200	13	18	22	27
1400	12	17	21	26
1600	12	16	20	24
1800	11	15	19	23

FORMS—CONCRETE (WALE SPACING)

Concrete Pressure (Lbs./ Square Foot)	Approx. Wale Spacing (in Inches) with Stud/Sheathing Combinations of:				
200	48	33	—	—	—
400	32	24	—	54	45
600	24	18	57	44	36
800	20	14	48	36	30
1000	17	12	42	31	26
1200	15	11	38	27	23
1400	14	10	33	24	20
1600	12	9	30	22	19
1800	—	—	27	20	17

FORMS—CONCRETE SLAB (STEEL RIB)

Span Lengths		
Thickness of Concrete	**Clear Span**	
	20 – Ga.	**18 – Ga.**
2 1/2"	7' 0"	8' 0"
3"	6' 6"	7' 6"
3 1/2"	6' 6"	7' 0"
4"	6' 0"	7' 0"
4 1/2"	6' 0"	6' 6"
5"	5' 6"	6' 6"
5 1/2"	5' 6"	6' 0"
6"	5' 6"	6' 0"
6 1/2"	5' 0"	5' 6"
7"	5' 0"	5' 6"

Check local codes

FORMS—ROUND CONCRETE, PAPER, FIBER
(SIZES, STRENGTHS, WEIGHTS)

Inside Dia. (In.)	Thickness of Walls (In.)	Inside to Outside Bursting Pressure (P.S.I.)	Wt./1000 Lin. Ft. (Lbs.)
3	.150	274	485
4	.175	233	740
5	.175	188	920
6	.200	188	1,275
7	.200	173	1,440
8	.200	143	1,635
9	.200	129	1,890
10	.225	127	2,280
11	.225	119	2,490
11 1/2	.225	122	2,600
12	.225	112	2,720
13 1/2	.225	111	3,050
14	.250	121	3,490
15	.300	149	4,450
16	.300	131	4,740
18	.300	111	5,325
20	.375	115	7,325
22	.375	112	8,000
24	.375	91	8,760
26	.375	78	9,480
28	.375	75	10,200
30	.400	85	11,670
32	.400	77	12,380
34	.400	71	13,180
36	.410	78	14,250
38	.450	61	16,800
40	.450	55	17,680
42	.450	60	18,550
44	.475	61	20,465
46	.475	56	21,380
48	.500	64	23,415

FORMULAS

Areas

AREA OF:

Circle = half diameter × half circumference.
Circle = square of diameter × .7854.
Circle = square of circumference × .07958.
Circular ring = sum of the diameter of the two circles × difference of the diameter of the two circles and that product × .7854.
Ellipse = product of the two diameters × .7854.
Parabola = base × ²/₃ of altitude.
Parallelogram = base × altitude.
Rectangle = length × breadth or height.
Regular polygon = sum of its sides × perpendicular from its center to one of its sides divided by 2.
Sector of circle = length of arc × ¹/₂ radius.
Segment of circle = area of sector of equal radius - area of triangle, when the segment is less, and plus area of triangle, when segment is greater than the semi-circle.
Square = length × breadth or height.
Top of a tank = multiply the square of the diameter by .7854.
Trapezium = divide into two triangles, total their areas.
Trapezoid = altitude × ¹/₂ the sum of parallel sides.
Walls of a tank = multiply the height by the circumference.

Circumference

Circle = diameter × 3.1416.
Circle = radius × 6.283185.

Sphere = square root of surface × 1.772454.
Sphere = cube root of solidity × 3.8978.

Contents (of Solids)

Cube = height × width × depth.
Frustum of pyramid or cone = sum of circumference at both ends × ¹/₂ slant height plus area of both ends.
Frustum of pyramid or cone = multiply areas of two ends together and extract square foot. Add to this root the two areas and × ¹/₃ altitude.
Wedge = area of base × ¹/₂ altitude.

Diameter

– of a circle = circumference × .3183.
– of a circle = square root of area × 1.12838.
– of a circle that shall contain area of a given square = side of square × 1.1284.
– of a sphere = cube root of solidity × 1.2407.
– of a sphere = square root of surface × .56419.

Formulas

A circle = circumference × .0159155.
A circle = ¹/₂ diameter.

Formulas continued.

Spheres

Solidity = surface × 1/6 diameter.
Solidity = cube of diameter × .5236.
Solidity = cube of radius × 4.1888.
Solidity = cube of circumference × .016887.

Contents of segment of sphere = (height squared plus three times the square of radius of base) × (height × .5236).
Contents of a sphere = diameter × .5236.

Side of inscribed cube of sphere = radius × 1.1547.
Side of inscribed cube of sphere = square root of diameter.

Areas of spheres
Multiply the square of the diameter by 3.1416.

Table of Spheres:

Dia.	Approx. No. of Sq. Ft.	Dia.	Approx. No. of Sq. Ft.
1	3.2	10	314.2
2	12.6	11	380.2
3	28.3	12	452.4
4	50.3	15	707.0
5	78.6	20	1256.6
6	113.1	25	1963.5
7	154.0	30	2827.0
8	201.0	40	5026.0
9	254.5		

Squares

A side multiplied by 1.4142 equals diameter of its circumscribing circle.
A side multiplied by 4.443 equals circumference of its circumscribing circle.
A side multiplied by 1.128 equals circumference of an equal circle.
A side multiplied by 3.547 equals circumference of an equal circle.

Side of inscribed equilateral triangle = diameter × .86.
Side of inscribed square = diameter × .7071.
Side of inscribed square = circumference × .225.
Side of square that shall equal area of circle = diameter × .8862.
Side of square that shall equal area of circle = circumference × .2821.

Surface Areas

Surface of pyramid or cone = circumference of base × 1/2 of the slant height plus area of base.
Surface of cylinder or prism = area of both ends plus length and × circumference.
Surface of sphere = diameter × circumference.

Volume of Solids

Volume of a Cone = Area of the base × 1/3 the height at center.
Volume of a Pyramid = Height × width × depth.
Volume of a Cube
Volume of a Cylinder = Area of the base × the height.
Volume of a Sphere = Diameter cubed × .5236.

FUELS—HEAT CONTENT IN BTUS

Fuel	Btu	Unit
Coal, anthracite	25,400,000	Ton
Coal, bituminous	26,200,000	Ton
Coal, lignite	14,000,000	Ton
Coal, briquettes	28,000,000	Ton
Electricity	3,412	kWh
Gas, butane	4,284,000	Bbl.
Gas, natural	1,035,000	1000 Cu. Ft.
Gas, natural	1,000	Therm
Gas, propane	3,843,000	Bbl.
Petroleum, crude oil	5,800,000	Bbl.
Petroleum, diesel	5,806,000	Bbl.
Petroleum, gasoline	5,253,000	Bbl.
Petroleum, kerosene	5,670,000	Bbl.
Wood, ash (dry)	20,000,000	Cord
Wood, aspen (dry)	12,500,000	Cord
Wood, beech (dry)	21,800,000	Cord
Wood, red maple (dry)	18,600,000	Cord
Wood, red oak (dry)	21,300,000	Cord
Wood, white pine (dry)	12,100,000	Cord

G

GAUGE—SHEET METAL

| STEEL SHEETS | | | GALVANIZED SHEETS | | STAINLESS STEEL SHEETS | | | |

Gauge	Thick-ness (In.)	Wt./ Sq. Ft. (Lbs.)	Gal-vanized Sheet (Gauge)	Wt. per Sq. Ft. (Lbs.)	Stain-less Sheet (Gauge)	Thick-ness (In.)	Wt./Sq. Ft. (Lbs.) Straight Chrome Alloys	Chrome Nickel Alloys
7	.1793	7.500						
8	.1644	6.875						
9	.1494	6.250						
10	.1345	5.625	10	5.7812	10	.140625	5.7937	5.9062
11	.1196	5.000	11	5.1562	11	.125000	5.1500	5.2500
12	.1046	4.375	12	4.5312	12	.109375	4.5063	4.5937
13	.0897	3.750	13	3.9062	13	.093750	3.8625	3.9375
14	.0747	3.125	14	3.2812	14	.078125	3.2187	3.2812
15	.0673	2.812	15	2.9687	15	.070312	2.8968	2.9531
16	.0598	2.500	16	2.6562	16	.062500	2.5750	2.6250
17	.0538	2.250	17	2.4062	17	.056250	2.3175	2.3625
18	.0478	2.000	18	2.1562	18	.050000	2.0600	2.1000
19	.0418	1.750	19	1.9062	19	.043750	1.8025	1.8375
20	.0359	1.500	20	1.6562	20	.037500	1.5450	1.5750
21	.0329	1.375	21	1.5312	21	.034375	1.4160	1.4437
22	.0299	1.250	22	1.4062	22	.031250	1.2875	1.3125
23	.0269	1.125	23	1.2812	23	.028125	1.1587	1.1813
24	.0239	1.000	24	1.1562	24	.025000	1.0300	1.0500
25	.0209	.875	25	1.0312	25	.021875	.9013	.9187
26	.0179	.750	26	.9062	26	.018750	.7725	.7875
27	.0164	.6875	27	.8437	27	.017187	.7081	.7218
28	.0149	.625	28	.7812	28	.015625	.6438	.6562
29	.0135	.5625	29	.7187	29	.014062	.5794	.5906
30	.0120	.500	30	.6562	30	.012500	.5150	.5250

GAUGE—WIRE

AWG (American Wire Gauge) or B & S (Brown & Sharpe)

Gauge	Diameter (In.)
000000	0.5800
00000	0.5165
0000	0.4600
000	0.4096
00	0.3648
0	0.3249
1	0.2893
2	0.2576
3	0.2294
4	0.2043
5	0.1819
6	0.1620
7	0.1443
8	0.1285
9	0.1144
10	0.1019
11	0.0907
12	0.0808
13	0.0720
14	0.0641
15	0.0570
16	0.0508
17	0.0453
18	0.0403
19	0.0359
20	0.0320
21	0.0285
22	0.0254
23	0.0226
24	0.0201
25	0.0179
26	0.0159
27	0.0142
28	0.0126
29	0.0113
30	0.0100
31	0.0089
32	0.0080
33	0.0071
34	0.0063
35	0.0056
36	0.0050
37	0.0045
38	0.0040
39	0.0035
40	0.0031

GAUGE—WIRE

Steel Wire Gauge (SWG or StlWG) Also American Steel and Wire Co. Gauge, Roebling Gauge, and Washburn and Moen Gauge, but not to be Confused with British Standard Wire Gauge.

Gauge No.	Diameter (In.)
0000	.3938
000	.3625
00	.3310
0	.3065
1	.2830
2	.2625
3	.2437
4	.2253
5	.2070
6	.1920
7	.1770
8	.1620
9	.1483
10	.1350
11	.1205
12	.1055
13	.0915
14	.0800
15	.0720
16	.0625
17	.0540
18	.0475
19	.0410
20	.0348
21	.0317
22	.0286
23	.0258
24	.0230
25	.0204
26	.0181
27	.0173
28	.0162
29	.0150
30	.0140

GIRDERS—BUILT-UP WOOD
Sizes for Various Loads and Spans
(Based on Douglas Fir 4-Square Guideline Framing)
Deflection Not Over 1/360 of Span—
Allowable Fiber Stress 1600 Pounds/Square Inch

Load/ Linear Foot of Girder	Length of Span				
	6'-0"	7'-0"	8'-0"	9'-0"	10'-0"
	Nominal Size of Girder Required (In.)				
750	6×8	6×8	6×8	6×10	6×10
900	6×8	6×8	6×10	6×10	8×10
1050	6×8	6×10	8×10	8×10	8×12
1200	6×10	8×10	8×10	8×10	8×12
1350	6×10	8×10	8×10	8×12	10×12
1500	8×10	8×10	8×12	10×12	10×12
1650	8×10	8×12	10×12	10×12	10×14
1800	8×10	8×12	10×12	10×12	10×14
1950	8×12	10×12	10×12	10×14	12×14
2100	8×12	10×12	10×14	12×14	12×14
2250	10×12	10×12	10×14	12×14	12×14
2400	10×12	10×14	10×14	12×14	
2550	10×12	10×14	12×14	12×14	
2700	10×12	10×14	12×14		
2850	10×14	12×14	12×14		
3000	10×14	12×14			
3150	10×14	12×14			
3300	12×14	12×14			

- The 6-in. girder is figured as being made with three pieces 2 in. dressed to 1 5/8 in. thickness
- The 8-in. girder is figured as being made with four pieces 2 in. dressed to 1 5/8 in. thickness.
- The 10-in. girder is figured as being made with five pieces 2-in. dressed to 1 5/8 in. thickness.
- The 12-in. girder is figured as being made with six pieces 2 in. dressed to 1 5/8 in. thickness.
- Note—For solid girders multiply above loads by 1.130 when 6-inch girder is used; 1.150 when 8-in. girder is used; 1.170 when 10-in. girder is used and 1.180 when 12-in. girder is used.

GLASS—BASIC FLAT (PROPERTIES)

Type	Thickness (In.)	Weight (Lbs./Sq. Ft.)	Light Size (In.)	Transmission (%)
Window	3/32 (SS)*	1.22	120 ui**	91
	1/8 (DS)***	1.63	140 ui**	91
Heavy Sheet	3/16	2.50	84 × 120	90
	7/32	2.85	84 × 120	89
Plate or Float	1/8	1.64	74 × 120	90
	1/4	3.28	128 × 204	89
Heavy Plate or Float	5/16	4.10	124 × 200	87
	3/8	4.92	124 × 200	86
	1/2	6.54	120 × 200	84
	5/8	8.17	120 × 200	82
	3/4	9.18	115 × 200	81
	7/8	11.45	115 × 200	79

*SS = Single Strength
**ui = united inches, the total of length plus width
***DS = Double Strength

GLASS—BLOCK
(Dimensions, set in wood)
12-INCH BLOCK   8-INCH BLOCK

No. of Block	Opening Width	Opening Height	No. of Block	Opening Width	Opening Height
1	1′ – 1 3/4″	1′ – 1 3/16″	1	0 3/4″	
2	2′ – 1 3/4″	2′ – 1 3/16″	2	1′ – 5 3/4″	1′ – 5 3/16″
3	3′ – 1 13/16″	3′ – 1 1/4″	3	2′ – 1 3/4″	2′ – 1 3/16″
4	4′ – 1 13/16″	4′ – 1 1/4″	4	2′ – 9 3/4″	2′ – 9 3/16″
5	5′ – 1 13/16″	5′ – 1 1/4″	5	3′ – 5 3/4″	3′ – 5 3/16″
6	6′ – 1 7/8″	6′ – 1 5/16″	6	4′ – 1 3/4″	4′ – 1 3/16″
7	7′ – 1 7/8″	7′ – 1 5/16″	7	4′ – 9 3/4″	4′ – 9 3/16″
8	8′ – 1 7/8″	8′ – 1 5/16″	8	5′ – 5 3/4″	5′ – 5 3/16″
9	9′ – 1 15/16″	9′ – 1 3/8″	9	6′ – 1 3/4″	6′ – 1 3/16″
10	10′ – 2″	10′ – 1 3/8″	10	6′ – 9 3/4″	6′ – 9 3/16″
			11	7′ – 5 3/4″	7′ – 5 3/16″
			12	8′ – 1 3/4″	8′ – 1 3/16″
			13	8′ – 9 3/4″	8′ – 9 3/16″
			14	9′ – 5 3/4″	9′ – 5 3/16″
			15	10′ – 1 3/4″	10′ – 1 3/16″

Dimensions given are frame opening dimensions and include clearance at each side and top of panel for wedging. Panel limits - 75 sq. ft. - 10 ft. in width.

GLASS—BLOCK
Layout Dimension 1/4-Inch Joints. All Blocks 3 7/8 Inch Thick

5 3/4-INCH SQUARE BLOCKS (6 INCH)

No. of Units	Panel Width or Height	Opening Width	Opening Height
1	5 3/4"	6 1/2"	6 3/8"
2	11 3/4"	1'-0 1/2"	1'-0 3/8"
3	1'-5 3/4"	1'-6 1/2"	1'-6 3/8"
4	1'-11 3/4"	2'-0 1/2"	2'-0 3/8"
5	2'-5 3/4"	2'-6 1/2"	2'-6 3/8"
6	2'-11 3/4"	3'-0 1/2"	3'-0 3/8"
7	3'-5 3/4"	3'-6 1/2"	3'-6 3/8"
8	3'-11 3/4"	4'-0 1/2"	4'-0 3/8"
9	4'-5 3/4"	4'-6 1/2"	4'-6 3/8"
10	4'-11 3/4"	5'-0 1/2"	5'-0 3/8"
11	5'-5 3/4"	5'-6 1/2"	5'-6 3/8"
12	5'-11 3/4"	6'-0 1/2"	6'-0 3/8"
13	6'-5 3/4"	6'-6 1/2"	6'-6 3/8"
14	6'-11 3/4"	7'-0 1/2"	7'-0 3/8"
15	7'-5 3/4"	7'-6 1/2"	7'-6 3/8"
16	7'-11 3/4"	8'-0 1/2"	8'-0 3/8"
17	8'-5 3/4"	8'-6 1/2"	8'-6 3/8"
18	8'-11 3/4"	9'-0 1/2"	9'-0 3/8"
19	9'-5 3/4"	9'-6 1/2"	9'-6 3/8"
20	9'-11 3/4"	10'-0 1/2"	10'-0 3/8"

7 3/4-INCH SQUARE BLOCKS (8 INCH)

No. of Units	Panel Width or Height	Opening Width	Opening Height
1	7 3/4"	8 1/2"	8 3/8"
2	1'-3 3/4"	1'-4 1/2"	1'-4 3/8"
3	1'-11 3/4"	2'-0 1/2"	2'-0 3/8"
4	2'-7 3/4"	2'-8 1/2"	2'-8 3/8"
5	3'-3 3/4"	3'-4 1/2"	3'-4 3/8"
6	3'-11 3/4"	4'-0 1/2"	4'-0 3/8"
7	4'-7 3/4"	4'-8 1/2"	4'-8 3/8"
8	5'-3 3/4"	5'-4 1/2"	5'-4 3/8"
9	5'-11 3/4"	6'-0 1/2"	6'-0 3/8"
10	6'-7 3/4"	6'-8 1/2"	6'-8 3/8"
11	7'-3 3/4"	7'-4 1/2"	7'-4 3/8"
12	7'-11 3/4"	8'-0 1/2"	8'-0 3/8"
13	8'-7 3/4"	8'-8 1/2"	8'-8 3/8"
14	9'-3 3/4"	9'-4 1/2"	9'-4 3/8"
15	9'-11 3/4"	10'-0 1/2"	10'-0 3/8"

11 3/4-INCH SQUARE BLOCKS (12 INCH)

No. of Units	Panel Width or Height	Opening Width	Opening Height
1	11 3/4"	1'-0 1/2"	1'-0 3/8"
2	1'-11 3/4"	2'-0 1/2"	2'-0 3/8"
3	2'-11 3/4"	3'-0 1/2"	3'-0 3/8"
4	3'-11 3/4"	4'-0 1/2"	4'-0 3/8"
5	4'-11 3/4"	5'-0 1/2"	5'-0 3/8"
6	5'-11 3/4"	6'-0 1/2"	6'-0 3/8"
7	6'-11 3/4"	7'-0 1/2"	7'-0 3/8"
8	7'-11 3/4"	8'-0 1/2"	8'-0 3/8"
9	8'-11 3/4"	9'-0 1/2"	9'-0 3/8"
10	9'-11 3/4"	10'-0 1/2"	10'-0 3/8"

Maximum panel size depends upon construction details

GLASS—BLOCK (MORTAR MIX)

Cement . . . . 1/4 bag
Lime . . . . 1/4 bag
Sand. . . . . 1 cu. ft.
Waterproofing Compound . . . . 1/2 pint

GLASS—BLOCK (MORTAR REQUIREMENTS)
Mortar for 1,000 Glass Block—1/4-inch Joints, All 3 7/8 Inches Thick

Size of Units	Mortar (Cu. Ft.)
5 3/4 × 5 3/4	7 1/8
7 3/4 × 7 3/4	9 1/2
11 3/4 × 11 3/4	14 5/8

GLASS—BLOCK (MORTAR REQUIREMENTS)
Per 100 Square Feet of Wall—1/4-inch Joints

Size of Block	5 3/4″	7 3/4″	11 3/4″
Number of Blocks	400	225	100
Weight of Panel	2000 lb.	1800 lb.	1900 lb.
Volume of Mortar	4.3 cu. ft.	3.2 cu. ft.	2.2 cu. ft.

GLASS—BLOCK (SIZES, WEIGHTS)

Sizes (In.)	Wt. Ea. (Lbs.)
5 3/4 × 5 3/4	3 1/2
7 3/4 × 7 3/4	6
11 3/4 × 11 3/4	16

Weights vary depending upon type of block

GLASS—INSULATING (PROPERTIES)

Type	Glass (In. Thick.)	Space (In.)	Total Thickness	Weight (Lbs./Sq. Ft.)	Size (Sq. Ft.)
Metal—edge	1/8 – 1/8	1/4	1/2	3.27	22
	1/8 – 1/8	1/2	3/4	3.27	22
	3/16 – 3/16	1/4	5/8	4.90	34
	3/16 – 3/16	1/2	7/8	4.90	42
	1/4 – 1/4	1/4	3/4	6.50	60
Glass—edge	3/32 – 3/32	3/16	3/8	2.40	10
	1/8 – 1/8	3/16	7/16	3.20	24

GLASS—LAMINATED, WITH PLASTIC INTERFACE (PROPERTIES)

Type	Thickness (In.)	Weight (Lbs./Sq. Ft.)	Area (In.)
3/32 – 3/32 sheet	13/64	2.45	48 × 80
7/64 – 7/64 sheet	15/64	2.90	48 × 100
1/8 – 1/8 sheet	1/4	3.30	48 × 100
1/8 – 1/8 float	1/4	3.30	72 × 120
3/16 – 3/16 sheet	3/8	5.00	72 × 120
1/4 – 1/4 float	1/2	6.56	72 × 120
5/16 – 5/16 float	5/8	8.20	72 × 120
3/8 – 3/8 float	3/4	9.84	72 × 120
7/16 – 7/16 float	7/8	11.40	72 × 120
1/2 – 1/2 float	1	13.08	72 × 120

Maximum size varies with the manufacturer. Also available in acoustical, bullet-resistant, and burglar-resistant types, and with two or more plastic interfaces.

GLASS—PATTERNED (PROPERTIES)

Thickness (In.)	Weight (Lbs./Sq. Ft.)	Size (In.)	Light Transmission (%)
1/8	1.60 – 2.10	60 × 132	80 – 90
7/32	2.40 – 3.00	60 × 132	80 – 90

Both thicknesses are available in floral, fluted, granular, hammered, ribbed, stippled, and striped types. Other proprietary patterns available.

GLASS—REFLECTIVE COATED (PROPERTIES)

- Type: Plate or float glass coated with a transparent, reflective metal or metallic oxide film.
- Thicknesses (inches): $1/8$, $1/4$.
- Weights: The same as for basic flat glass.
- Light transmission % (from outside): Range 8 to 50.
- Reflectance % (outside, daylight): Range 6 to 45.
- Sizes (inches): 66×114 to 117×140.
- Colors: Blue, blue-green, bronze, copper, gold, neutral, silver. Tones variable; other colors sometimes available.
- Applications: General glazing, architectural, spandrel.
- Styles: Available in custom shapes and sizes, other thicknesses, in combinations with clear or tinted glass, in strengthened form, and in insulating glazing.

GLASS—STRENGTHENED (PROPERTIES)

- Types: Heat-strengthened, chemically strengthened, fully tempered.
- Thicknesses (inches): $1/8$, $5/32$, $3/16$, $1/4$, $5/16$, $3/8$, $1/2$, $5/8$, $3/4$, $7/8$.
- Weights: Approximately the same as for basic flat glass.
- Light transmission %: The same as for basic flat glass.
- Sizes (inches): Range from approximately 42×76 to 96×132, depending upon manufacturer, type of glass, and application.
- Styles: Available in custom shapes and sizes, some tinted types, some reflective coated types, and insulating glazing form.

GLASS—TINTED (PROPERTIES)

Tint	Thickness (In.)	Solar Transmission (%)	Light Transmission (%)
Bronze	$1/8$	65	68
	$3/16$	55	58
	$1/4$	46	50
	$3/8$	33	37
	$1/2$	24	28
Gray	$1/8$	63	62
	$3/16$	53	51
	$1/4$	44	42
	$3/8$	31	28
	$1/2$	24	20
Green	$3/16$	55	78
	$1/4$	48	74

Sizes and weights are approximately the same as for basic flat glass with the exception of green tints; the weights are the same but the sizes are generally smaller. Tints vary, as do the glass properties; blue-green is also available.

GLASS—WIND RESISTANCE
Maximum Square Feet of Area Required

Glass Thickness	SS	DS	1/8"	3/16"	1/4"	5/16"	3/8"	1/2"	5/8" up to 1"	1 1/4"
30 Mile Wind	35 sq. ft.	64.5	72	162	288	244				
40 Mile Wind	17.5	32.25	36	81	144	225	248			
55 Mile Wind	11.6	21.5	24	54	96	150	216			
65 Mile Wind	8.7	16.1	18	41	72	112	162	244		
80 Mile Wind	5.8	10.85	12	27	48	75	108	192		
100 Mile Wind	3.5	6.45	7	16	29	45	65	115		
120 Mile Wind	2.5	4.6	5	11	20	32	46	82	85	81

GLASS—WIRED (PROPERTIES)

Type	Thickness (In.)	Weight (Lbs./Sq. In.)	Size (In.)	Light Transmission (%)
Parallel wired	7/32	2.82	54 × 120	80 – 85
	1/4	3.50	60 × 144	80 – 85
	3/8	4.45	60 × 144	80 – 85
Patterned— (diamond mesh, square mesh) Polished— (diamond mesh, square mesh)	1/4	3.40	60 × 144	80 – 85

GLAZING COMPOUND

- Packages: 5, 10, 25, 50, 100 pounds
- 1 gallon will glaze: 150 linear feet of sash

GROUT—QUANTITIES REQUIRED

Types	Nominal Sizes (In.)	Joint Width (In.)	Sq. Ft./Bag	
			25 (Lb.)	50 (Lb.)
Pavers	4× 4× 1/2	1/4	45	90
	4× 4× 3/8	3/8	30	60
Tile	6× 6× 1/2	1/4	60	120
	6× 6× 1/2	3/8	45	90
Tile	6× 6× 3/4	1/4	42	84
	6× 6× 3/4	3/8	30	60
Tile	8× 8× 1/2	1/4	50	100
	8× 8× 3/4	3/8	55	110
Tile	9× 9× 3/4	1/4	65	130
	9× 9×1 3/4	3/8	70	140
Tile	12×12× 3/4	1/2	50	100
Brick	4× 8× 3/4	1/4	40	80
	4× 8×1 1/2	1/4	55	110
Mosaic	1× 1× 1/4	1/16	120	240
	1× 2× 1/4	1/16	150	300
Wall	4× 4× 5/16	1/16	300	600

Estimated quantities based on full depth joints, no waste. Adjust for other sizes, types, joints, and existing or unusual wall or base conditions.

GUTTERS—FIR (SIZES, WEIGHTS)

Size (In.)	Finished Size (In.)	Lengths	Av. Wt. Lin. Ft. (Lbs.)
3×4	2 1/2×3 1/2	Up to 20′	1 1/2
3×5	2 1/2×4 1/2	Up to 20′	1 1/2
4×4	3 1/2×3 1/2	Up to 20′	1 1/2
4×5	3 1/2×4 1/2	Up to 20′	1 3/4
4×6	3 1/2×5 1/2	Up to 20′	2
5×7	4 1/2×6 1/2	Up to 20′	3

GUTTERS—GALVANIZED METAL

Depth (In.)	Width (In.)	Gauge	Lengths
3 1/2	3 1/2	28 – 29	10
4 1/4	4 3/8		10
5	5		10

GUTTERS—GALVANIZED STEEL
(LAP & SLIP SEAL JOINTS)

Sizes (In.)	Lengths	Gauge
3 1/2	10	26 – 29
4	10	26 – 29
5	10	26 – 29
6	10	26 – 29

HARDBOARD

Nominal Thickness (In.)	Thickness Tolerance: Min. – Max. (In.)
$1/12$ (0.083)	0.070 – 0.090
$1/10$ (.100)	.091 – .110
$1/8$ (.125)	.115 – .155
$3/16$ (.188)	.165 – .205
$1/4$ (.250)	.210 – .265
$5/16$ (.312)	.290 – .335
$3/8$ (.375)	.350 – .400
$7/16$ (.438)	.410 – .460
$1/2$ (.500)	.475 – .525
$5/8$ (.625)	.600 – .650
$11/16$ (.688)	.660 – .710
$3/4$ (.750)	.725 – .775
$13/16$ (.812)	.785 – .835
$7/8$ (.875)	.850 – .900
1 (1.000)	.975 – 1.025
1 $1/8$ (1.125)	1.115 – 1.155

Hardboard—Classifications, see pp. 108-109.

HARDWOOD—WEIGHTS
Per 1000 Feet, ⁴/₄ and Thicker

(EASTERN) WOODS

	Dry	Green		Dry	Green
Ash	3,800	4,800	Gum, Black	3,500	5,200
Basswood	2,500	4,200	Hickory	4,500	6,100
Beech	4,000	5,700	Maple, Hard	4,000	5,500
Birch	4,000	5,700	Maple, Soft	3,600	5,000
Buckeye	2,500	4,000	Oak, Red	4,000	5,500
Butternut	2,800	4,000	Oak, White	4,000	5,500
Cherry	3,900	5,000	Poplar	2,800	4,000
Chestnut	2,800	5,000	Walnut	3,800	4,900
Elm	4,000	5,500			

NORTHERN

	Dry	Green		Dry	Green
Ash, Brown or Black	3,300	4,600	Elm, Rock	4,000	5,400
Aspen (Popple)	2,800	4,500	Elm, Soft	3,200	5,000
Basswood	2,500	4,200	Maple Hard	4,300	5,500
Beech	4,000	5,700	Maple, Soft	3,700	5,000
Birch	4,000	5,500	Oak, Red	4,000	5,000

NORTHWESTERN

	Dry	Green
Alder, Red	3,200	5,000

SOUTHERN

	Dry	Green		Dry	Green
Ash (Firm Tex)	3,900	4,600	Hackberry	3,300	4,600
Ash (Soft Tex)	3,600	4,600	Hickory	4,500	6,000
Basswood	2,600	4,200	Magnolia	3,300	4,700
Beech	4,100	5,700	Maple, Soft	3,100	5,000
Cedar,			Oak, Red	4,200	6,000
Aromatic Red	3,300	3,800	Oak, White	4,200	6,200
Cottonwood	2,800	4,600	Pecan	4,300	5,900
Cypress	2,900	5,000	Poplar	3,000	4,000
Elm, Rock	4,000	5,400	Sycamore	3,400	4,800
Elm, Soft	3,300	5,000	Tupelo	3,400	5,100
Gum, Black	3,400	5,100	Walnut	3,800	4,900
Gum, Red	3,500	5,500	Willow	2,700	4,200
Gum, Sap	3,300	5,200			

HARDBOARD—CLASSIFICATIONS

Class	Nominal Thickness (In.)	Water Resistance (Max Av Per Panel)		Modulus of Rupture: Min. Av. per Panel (psi)	Tensile Strength (Min. Av. per Panel)	
		Water Absorption Based on Weight (Percent)	Thickness Swelling (Percent)		Parallel to Surface (psi)	Perpendicular to Surface (psi)
Tempered	1/12	30	25	6000	3000	130
	1/10 1/8 3/16	25	20			
	1/4 5/16 3/8	20 15 10	15 10 9			
Standard	1/12	40	30	4500	2200	90
	1/10 1/8 3/16	35	25			
	1/4 5/16 3/8	25 20 15	20 15 10			

Service-Tempered	1/8	35	30	4500	2000	75
	3/16	30	30			
	1/4	30	25			
	3/8	20	15			
Service	1/8	45	35	3000	1500	50
	3/16	40	35			
	1/4	40	30			
	3/8	35	25			
	7/16	35	25			
	1/2	30	20			
	5/8	25	20			
	11/16	25	20			
	3/4	20	15			
	13/16					
	7/8					
	1					
	1 1/8					
Industrialite	1/4	50	30	2000	1000	25
	3/8	40	25			
	7/16	40	25			
	1/2	35	25			
	5/8	30	20			
	11/16	30	20			
	3/4	25	20			
	13/16					
	7/8					
	1					
	1 1/8					

HINGES—CLEARANCE OF BUTT HINGES

Door Thickness (In.)	Size of Butt (In.)	Max. Clearance (In.)
1 3/8	3 × 3	3/4
	3 1/2 × 3 1/2	7/8
	4 × 4	1 5/8
1 3/4	4 × 4	1
	4 1/2 × 4 1/2	1 3/8
	5 × 5	2
2	4 1/2 × 4 1/2	1 1/4
	5 × 5	1 3/4
	6 × 6	2 3/4
2 1/4	5 × 5	1 1/4
	6 × 6	2 1/4
	6 × 7	3 1/4
	6 × 8	4 1/4

HINGES—RECOMMENDED NUMBER PER DOOR

Doors 60 inches high and under . . . . 2 butt hinges
Doors over 60 inches high and not over 90 inches high . . . . 3 butt hinges
Doors over 90 inches high and not over 120 inches high . . . . 4 butt hinges
Transoms 48 inches wide and under . . . . 2 butt hinges
Transoms over 48 inches wide and not over 84 inches wide . . . . 3 butt hinges

HINGES—SIZES

Thickness of Doors (In.)	Width of Doors or Height of Transoms (In.)	Height of Butt Hinges: Length of Joint (In.)	
3/4 and 7/8 cupboard doors	To 24	2 1/2	
7/8 and 1 1/8 screen doors	To 36	3	
1 1/8 doors	To 36	3 1/2	
1 1/4 and 1 3/8 doors	To 32	3 1/2	
Do	Over 32 to 37	4	
1 9/16, 1 3/4 & 1 7/8 doors	To 32	4 1/2	
Do	Over 32 to 37	5	
Do	Over 37 to 43	5	Extra heavy
Do	Over 43 to 50	6	Extra heavy
2, 2 1/4 & 2 1/2 doors	To 43	5	Extra heavy
Do	Over 43 to 50	6	Extra heavy
1 1/4 & 1 3/8 transoms	To 20	2 1/2	
Do	Over 20 to 36	3	
1 1/2, 1 9/16, 1 3/4 & 1 7/8 transoms	To 20	3	
Do	Over 20 to 36	3 1/2	
2, 2 1/4 & 2 1/2 transoms	To 20	3 1/2	
Do	Over 20 to 36	4	

HOT WATER—REQUIREMENTS
Maximum Daily Requirement for Hot Water;
Gallons for 24 Hours Apartments and Private Dwellings Including Laundry

Rooms	Number of Bathrooms				
	1	2	3	4	5
1	60	—	—	—	—
2	70	—	—	—	—
3	80	—	—	—	—
4	90	120	—	—	—
5	100	140	—	—	—
6	120	160	200	—	—
7	140	180	220	—	—
8	160	200	240	250	—
9	180	220	260	275	—
10	200	240	280	300	—
11	—	260	300	340	—
12	—	280	325	380	450
13	—	300	350	420	500
14	—	—	375	460	550
15	—	—	400	500	600
16	—	—	—	540	650
17	—	—	—	580	700
18	—	—	—	620	750
19	—	—	—	—	800
20	—	—	—	—	850

I

INSULANTS—TYPES & PROPERTIES

Type	Rated R Value (In./Thickness)	Max. Service Temp. (°F.)	Density (Lbs./Cu. Ft.)
Cellulose (loose)	3.2	180	2.2 – 3.0
Fiberglass (blanket)	3.17	180	1.5 – 2.5
Fiberglass (loose)	2.2	180	0.6 – 2.5
Fiberglass (rigid)	4.4	180	4 – 9
Isocyanurate (rigid)	7.2	200	1.6 – 2.0
Mineral Wool (blanket)	3.17	500	1.5 – 4.0
Mineral Wool (loose)	3.1	500	1.5 – 2.5
Perlite (loose)	2.7	400	2 – 11
Phenolic (rigid)	8.3	300	2.5
Polystyrene, Expanded (rigid)	4.0	165	0.9 – 1.6
Polystyrene, Extruded (rigid)	5.0	165	1.6 – 2.0
Vermiculite (loose)	2.7	1000	4 – 10

INSULATION—CELLULAR GLASS

Type	Size (In.)	Thickness (In.)
Flat Block	12 × 18 18 × 24	1 1/2 2, 2 1/2 3, 3 1/2 4, 4 1/2 5
Board	24 × 48	1 1/2, 2 2 1/2, 3 3 1/2, 4
Tapered Block	18 × 24	1/16 / ft. taper 1/8 / ft. taper 1/4 / ft. taper 1/2 / ft. taper

Typical R value = 2.63/in. thickness.
Approximate density = 8.5 lbs./cu. ft.
Water vapor permeability = 0.0/inch.

INSULATION—MASONRY FILL
(EXPANDED PERLITE OR VERMICULITE)
Approximately 4 Cubic Foot Bags Required to Fill

Sq. Ft. of Wall Area	1" Cavity	2" Cavity	2 1/2" Cavity	Concrete Masonry Units	
				8"	12"
100	2	4	5	7	13
500	10	20	25	34	63
1,000	21	42	50	69	125
2,000	42	84	100	138	250
3,000	62	124	150	207	375
5,000	104	208	250	345	625
7,000	146	292	350	483	875
10,000	208	416	500	690	1250

INSULATION—RIGID (AREAS COVERED)

Number of Pieces	Size of Boards								
	4×4	4×6	4×7	4×8	4×9	4×10	4×12	2×4	2×8
1	16	24	28	32	36	40	48	8	16
2	32	48	56	64	72	80	96	16	32
3	48	72	84	96	108	120	144	24	48
4	64	96	112	128	144	160	192	32	64
5	80	120	140	160	180	200	240	40	80
6	96	144	168	192	216	240	288	48	96
7	112	168	196	224	252	280	336	56	112
8	128	192	224	256	288	320	384	64	128
9	144	216	252	288	324	360	432	72	144
10	160	240	280	320	360	400	480	80	160

INSULATION—VERMICULITE

Thickness (In.)	2	3	3 3/4	4	5	5 3/4
Coverage (Sq. Ft. per 4 ft. bag)	26	17	14	13	10	9

1000 sq. ft. ceiling including joist 3″ thick requires 60 4-ft. bags

J

JOISTS—BAR (SIZES, WEIGHTS)

Depth (In.)	Width (In.)	Types	Bearing Plate (In.)	Height at Ends (In.)	Wts. per Lin. Ft. (Lbs.)
8	3 1/2	2	3 × 5	2 1/2	3.31 – 4.08
10	3 1/2	3	3 × 5	2 1/2	4.02 – 4.74 – 5.78
12	3 1/2	2	3 × 5	2 1/2	4.94 – 5.73
12	3 1/2		4 × 5	2 1/2	6.85 – 7.92
14	3 1/2	1	4 × 5	2 1/2	6.97
14	4 1/2	2	4 × 5	2 1/2	8.32 – 9.43
16	4 1/2	2	4 × 5	2 1/2	8.42 – 9.85
18	4 1/2	2	4 × 6	2 1/2	8.80 – 9.90
20	4 1/2	1	4 × 6	2 1/2	9.90

Many other sizes available

JOISTS—CONCRETE (SIZES, WEIGHTS)

Height (In.)	Width (In.)	Length	Wt. (Lbs.)/ Linear Ft.
6	3	to 40	
8	3	to 40	20 – 28
10	3	to 40	24 – 32
12	3 1/2	to 40	32 – 36
14	4	to 40	36 – 40

JOISTS—SPACING (WOOD)
Average 1000 – 2000 Pounds Fiber Stress

Joist Size	Spacing (In.)	Joist Lengths (Weight Is Total Load)				
		40 Lbs. Sq. Ft.	60 Lbs. Sq. Ft.	80 Lbs. Sq. Ft.	100 Lbs. Sq. Ft.	Weight of Joist (Lbs.)/ Sq. Ft. of Area
2 × 4	12	8'–7"	7'	6'	5'–4"	1 1/2
	16	7'–2"	6'	5'–2"	4'–8"	1 1/4
	18	6'–11"	5'–8"		4'–4"	1
	24	6'		4'–3"	3'–9"	3/4
2 × 6	12	13'	10'–8"	9'–3"	8'–3"	2 1/2
	16	11'–4"	9'–3"	8'	7'–2"	2
	18	10'–8"	8'–9"	7'–7"	6'–9"	1 3/4
	24	9'–9"	7'–7"	6'–6"	5'–10"	1 1/4
2 × 8	12	17'–6"	14'–5"	12'–4"	11'	3 1/2
	16	15'–2"	12'–4"	10'–8"	9'–7"	2 1/2
	18	14'–3"	11'–8"	10'	9'	2 1/4
	24	12'–4"	10'	8'–9"	8'	1 3/4
2 × 10	12	22'	18'	15'–8"	14'	4 1/4
	16	19'–2"	15'–8"	13'–6"	12'	3 1/4
	18	18'	14'–9"	12'–9"	11'–6"	3
	24	15'–8"	12'–10"	11'	10'	2 3/4
2 × 12	12	26'–8"	21'–10"	19'	17'	5 1/4
	16	23'–3"	19'	16'–4"	14'–8"	4
	18	21'	18'	15'–5"	13'–10"	3 1/2
	24	19'	15'–5"	13'–5"	12'	2 1/2

Check local codes

Joists—Steel Hangers, see p. 117.

JOISTS—WOOD (MAXIMUM SPANS FOR CEILING JOISTS)
Maximum Spans for Ceiling Joists

Size of Joists (In.)	Spacing of Joists (In.)	White Pine	Norway Pine or Spruce	Hemlock	Douglas Fir	Long Leaf Yellow Pine
2 × 6	12	10'–4"	10'–11"	11'–6"	12'–0"	12'–6"
2 × 6	16	9'–1"	9'–6"	10'–0"	10'–6"	10'–10"
2 × 8	12	13'–9"	14'–6"	15'–3"	15'–11"	16'–6"
2 × 8	16	12'–0"	12'–8"	13'–4"	13'–11"	14'–5"

Check local codes

JOISTS—STEEL HANGERS

Size Joist (In.)	Size (In.)	Weight (Lbs.)
2×6	2×$^1/_8$	1.4
2×8	2×$^1/_8$	1.6
2×10	2×$^1/_8$	1.9
2×12	2×$^5/_{32}$	2.7
2×14	2×$^3/_{16}$	3.8
3×6	2×$^1/_8$	1.5
3×8	2×$^5/_{32}$	1.8
3×10	2×$^3/_{16}$	3.0
3×12	2 $^1/_2$×$^3/_{16}$	4.3
3×14	3×$^3/_{16}$	5.8
4×6	2×$^1/_8$	1.6
4×8	2×$^5/_{32}$	1.9
4×10	2×$^3/_{16}$	3.2
4×12	2 $^1/_2$×$^3/_{16}$	4.5
4×14	3×$^3/_{16}$	6.0
6×6	2 $^1/_4$×$^3/_{16}$	3.0
6×8	2 $^1/_2$×$^3/_{16}$	3.8
6×10	3×$^3/_{16}$	5.2
6×12	3×$^3/_{16}$	6.0
6×14	3×$^1/_4$	8.9
6×16	3×$^5/_{16}$	12.0
8×8	3×$^3/_{16}$	4.9
8×10	3×$^1/_4$	7.7
8×12	3×$^5/_{16}$	10.6
8×14	3×$^5/_{16}$	11.7
8×16	3 $^1/_2$×$^5/_{16}$	16.5

JOISTS—WOOD (MAXIMUM SPANS FOR FLOOR JOISTS)

Size	Spacing (In.)	Wt. (Lbs.)/ Sq. Ft.	Long Leaf Yellow Pine	Fir	Hemlock	Spruce	White Pine
2× 6	12	2 $^1/_2$	10'–4"	10'	9'–10"	9'–4"	8'–8"
2× 6	16	2	9'	8'–10"	8'–6"	8'	7'–8"
2× 8	12	3 $^1/_2$	13'–10"	13'–6"	12'–8"	12'	11'–8"
2× 8	16	2 $^1/_2$	12'–2"	11'–8"	11'	10'–8"	10'
2×10	12	4 $^1/_4$	17'–4"	16'–8"	16'	15'–6"	14'–8"
2×10	16	3 $^1/_4$	15'–2"	14'–8"	14'	13'–6"	12'–8"
2×12	12	5 $^1/_4$	20'–6"	20'	19'–6"	18'–6"	17'–6"
2×12	16	4	18'–6"	17'–6"	16'–10"	16'	15'–2"

Check local codes

JOISTS-WOOD (SPANS & LOADS)

Joists Spaced 16 Inches on Center Uniformly Loaded

Live Load in Pounds/Square Foot of Floor Area

Nominal Size of Joist Required

Length of Span	10 Lbs.		20 Lbs.		30 Lbs.		40 Lbs.		50 Lbs.	
	Douglas Fir	W. Coast Hemlock	Douglas Fir	W. Coast Hemlock	Douglas Fir	W. Coast Hemlock	Douglas Fir	W. Coast Hemlock	Douglas Fir	W. Coast Hemlock
8' or less	2×6 in.	2×6 in.	2×6 in.	2×6 in.	2×6 in.	2×6 in.	2×6 in.	2×6 in.	2×6 in.	2×6 in.
8'–6"	2×6	2×6	2×6	2×6	2×6	2×6	2×6	2×6	2×6	2×8
9'–0"	2×6	2×6	2×6	2×6	2×6	2×6	2×6	2×8	2×8	2×8
9'–6"	2×6	2×6	2×6	2×6	2×6	2×8	2×8	2×8	2×8	2×8
10'–0"	2×6	2×6	2×6	2×6	2×8	2×8	2×8	2×8	2×8	2×8
10'–6"	2×6	2×6	2×8	2×8	2×8	2×8	2×8	2×8	2×8	2×8
11'–0"	2×6	2×6	2×8	2×8	2×8	2×8	2×8	2×8	2×8	2×8
11'–6"	2×6	2×8	2×8	2×8	2×8	2×8	2×8	2×8	2×8	2×8
12'–0"	2×8	2×8	2×8	2×8	2×8	2×10	2×10	2×10	2×10	2×10
12'–6"	2×8	2×8	2×8	2×10	2×10	2×10	2×10	2×10	2×10	2×10
13'–0"	2×8	2×8	2×8	2×10	2×10	2×10	2×10	2×10	2×10	2×10
13'–6"	2×8	2×8	2×8	2×10	2×10	2×10	2×10	2×10	2×10	2×10
14'–0"	2×8	2×8	2×10	2×10	2×10	2×10	2×10	2×10	2×10	2×10
14'–6"	2×8	2×8	2×10	2×10	2×10	2×10	2×10	2×12	2×12	2×12
15'–0"	2×8	2×10	2×10	2×10	2×10	2×12	2×12	2×12	2×12	2×12
15'–6"	2×10	2×10	2×10	2×10	2×12	2×12	2×12	2×12	2×12	2×12
16'–0"	2×10	2×10	2×10	2×10	2×12	2×12	2×12	2×12	2×12	2×12
16'–6"	2×10	2×10	2×10	2×12	2×12	2×12	2×12	2×12	2×12	2×12
17'–0"	2×10	2×10	2×10	2×12	2×12	2×12	2×12	2×12	2×12	2×12
17'–6"	2×10	2×10	2×12	2×12	2×12	2×12	2×12		2×12	
18'–0"	2×10	2×10	2×12	2×12	2×12	2×12	2×12			
18'–6"	2×10	2×12	2×12	2×12	2×12		2×12			
19'–0"	2×10	2×12	2×12	2×12						
19'–6"	2×12	2×12	2×12							
20'–0"	2×12	2×12	2×12							

Check local codes

K

K FACTOR—VARIOUS MATERIALS

Material	K Factor	
Clay & Shale Brick (4″ thick)	5.00	
Concrete (sand & gravel)	10.00	
Cinder	5.00	
Expanded Shale or Clay	3.00	
Slag	2.00	
Expanded Perlite or Vermiculite, 35# density	.85	
Hemlock, Fir, Pine	.65 to	.85
Cypress, Redwood	.70 to	.80
Maple, Oak, Birch	1.00 to	1.25
Balsa	.60	
Sawdust	.40 to	.50
Rockwool - Wood Fiber	.25 to	.35
Expanded Perlite, Vermiculite	.35 to	.45
Cellulose (fluffy)	.25	
Cork (granulated)	.32	
Cotton (fireproofed)	.24	
Wood, Wool	.25	
Hair & Jute Mixed	.67	
Crepe Paper	.27	
Cone, Wood Fiberboards	.34	
Terrazzo, Tile, Stucco	10.00 to 12.00	

L

LATH-GYPSUM (SPACING OF SUPPORTS)

	Nailed	
Type Lath	Walls	Ceilings
3/8″ Plain or Insulating	16″	16″
3/8″ Perforated	16″	16″
1/2″ Plain	24″	24″

LATH—GYPSUM (TYPICAL PRODUCT SIZES)

Type	Thickness (In.)	Length (Ft.)	Width (In.)
Plain	3/8	4, 8, 12	16, 24
	1/2	4	16, 24
Perforated	3/8	4	16
	1/2	4	16
Foil-backed	3/8	4, 8	16
	1/2	4	16
Type X	3/8	4, 8	16
Edges of all types are rounded.			

LATH—METAL (CONSTRUCTION DATA)

Type	Weight (Lbs.)/ Square Yard	Sheet Size	Wood (In.)	Maximum Spacings			
				Vertical		Horizontal	
				Metal			
				Solid Partitions (In.)	Others (In.)	Wood or Concrete (In.)	Metal (In.)
Diamond Mesh	2.5	27″ × 96″	16	16	12	16	13
Diamond Mesh	3.4	27″ × 96″	16	16	16	16	13
Flat Rib	2.75	27″ × 96″	16	16	16	16	12
Flat Rib	3.4	27″ × 96″	19	24	19	19	19
3/8″ Rib	3.4	24″ × 96″	24	24	24	24	24
3/8″ Rib	4.0	24″ × 96″	24	24	24	24	24
	Per Sq. Ft.						
3/4″ Riblath	0.60	2′ × 8′					
	0.75	10′ × 12′					

LATH—METAL (CORNER BEADS)

Type	Size	Ga.	Wts. (Lbs.) 1000 Ft.	Lengths
Corner Lath	2″ × 2″		100	8′
	3″ × 3″		160	8′
Exp. Corner Bead	2 1/2″	26	200	8-9-10-12′
Flex Corner Bead	1 9/16″	26	180	8-9-10-12′
Bull Nose Corner Bead	3/4″ Rad.	26	250	8-9-10-12′

Other types and sizes available

LATH—METAL (SIZES, WEIGHTS)

Type	Size (In.)	Wts. (Lbs.)/ Sq. Yd.		Packed (Bdl.)
Diamond Mesh	27 × 96	2.5 - 3.4	5/16" Mesh	10 Shts. 20 Sq. Yds.
Self-furring Diamond Mesh	27 × 96	2.5 - 3.4	5/16" Furred 3/8"	10 Shts. 20 Sq. Yds.
Stucco Lath	48 × 99	1.8 - 3.6	1 3/8" Furred 3/8"	10 Shts. 36 2/3 Sq. Yds.
Flat Rib Lath	24 × 96	2.75 - 3.4	1/8" Rib	9 Shts. 16 Sq. Yds.
	27 × 96			10 Shts. 20 Sq. Yds.
3/8" Rib Lath	24 × 96	3.4 - 4		9 Shts. 16 Sq. Yds.
	27 × 96			10 Shts. 20 Sq. Yds.
3/4" Rib Lath	29 × 72	6 - 7 1/2 lbs. per C Sq. Ft.	3/4" Rib 3 5/8" O.C.	10 Shts. 145 Sq. Ft.
	29 × 96	6 - 7 1/2 lbs. per C Sq. Ft.	3/4" Rib 3 5/8" O.C.	10 Shts. 193 1/3 Sq. Ft.
	29 × 120	6 - 7 1/2 lbs. per C Sq. Ft.	3/4" Rib 3 5/8" O.C.	10 Shts. 241 2/3 Sq. Ft.
	29 × 144	6 - 7 1/2 lbs. per C Sq. Ft.	3/4" Rib 3 5/8" O.C.	10 Shts. 290 Sq. Ft.

Some special weights and finishes available.

LATH—METAL (SPACING SUPPORTS)

Lath Type	Min. Wt. of Lath (Lbs./Sq. Yd.)	Maximum Allowable Spacing of Supports				
		Vertical Supports			Horizontal Supports	
		Wood (In.)	Metal		Wood or Concrete (In.)	Metal (In.)
			Solid Partitions (In.)	Others (In.)		
Diamond Mesh (Flat Expanded)	2.5	16	16	12	0	0
	3.4	16	16	16	16	13
Flat Rib	2.75	16	16	16	16	12
	3.4	19	24	19	19	19
3/8" Rib	3.4	24	24	24	24	24
	4.0	24	24	24	24	24
Sheet Lath	4.5	24	24	24	24	24
V Stiffening Lath	3.3	24	—	24	19	19
Wire Fabric 2×2 16 ga.	1.	16	—	16	16	16

Check local codes

LATH—NAILS FOR ATTACHING
METAL LATH TO WOOD SUPPORTS

	Type	Gauge	Length (In.)	Nailing Estimate for 100 Sq. Yds. of Lath (Lbs.)
Partitions	4d Common	12 1/2	1 1/2	5
	6d Common	11 1/2	2	8 3/4
	Wire Staples	14	1	4 1/4
	Roofing Nail, with 7/16" head	11	1	5 1/4
Ceilings	Barbed Roofing Nail, with 7/16" head	11	1 1/2	8 3/4

LETTERS—PLASTIC
Average Stock Sizes (Inches)

Size	Depth	Width Stroke	Total Width
4	3/4	3/4	2 3/4
6	1	1 1/4	4
9	1 5/8	2 1/8	6
12	2 1/8	2 5/8	8
15	2 1/4	3	10
18	2 3/4	3 3/4	11
24	2 3/4	5 1/4	16
30	3 1/2	6 1/4	20
36	4	7	24
48	5 1/2	10	32

Many sizes and styles available.

LIME

Hydrated Lime

Hydrated lime weighs 40 lbs. per cu. ft.
A 50 lb. bag of hydrated lime makes 1.09 cu. ft. lime putty.
100 lbs. of hydrated lime makes 2.18 cu. ft. lime putty.
45.8 lbs. of hydrated lime makes 1 cu. ft. lime putty.
170 lbs. of hydrated lime makes the same quantity of putty
    as 100 lbs. of quicklime.
A cu. ft. of lime putty made from hydrate weighs about 83 lbs.
1 cu. ft. finish lime will cover 18 sq. yds. white coat finish.
A ton of hydrated lime makes approximately 45 to 50 cu. ft.
    putty.

Lime Putty

A cu. ft. of lime putty weighs approximately 80 lbs.
A cu. ft. of lime putty is made from 45.8 lbs. of hydrated lime.
A cu. ft. of lime putty is made from 27.3 lbs. of quicklime.
A 12-qt. bucket holds approximately 30 lbs. of lime putty.
A cu. ft. of lime putty is approximately 2.7 12-qt. buckets.
A cu. ft. of lime putty is approximately 6.5 No. 2 shovels (new).

Quicklime

Quicklime weights 55-60 lbs. per cu. ft.
27.3 lbs. of quicklime will make 1 cu. ft. of lime putty.
100 lbs. of quicklime will make 3.69 cu. ft. of lime putty.
100 lbs. of quicklime will make as much putty as 170 lbs. of
    hydrate.
A cu. ft. of putty made from quicklime weights slightly over 80 lbs.
1 ton of quicklime makes approximately 80 cu. ft. of putty.

LIME—FINISH COATS (QUANTITY REQUIRED)

	Lbs. Required/ 100 Sq. Yds.
White Smooth:	400 Finish Lime
	100 Plaster of Paris
White Textured:	300 Finish Lime
	100 Plaster of Paris
Sand Finish:	200 Finish Lime
	600 Sand
	10 to 30 Keenes, white portland or retarded gypsum

LIME—TYPES

Type	Container (Lbs.)	Yield/Ton
Double Hydrated	50	40 – 50 cu. ft. putty
Hydrated	50	40 – 50 cu. ft. putty
Quick Lime	60-80	80 – 100 cu. ft. putty
All-Purpose Hydrated	5-10-25-50	
Agricultural Hydrate	50	

LINTELS—BRICK REINFORCED (SPANS)

Wall Thickness (In.)	Lintel Height (In.)	Bars (In.)				
		$2^{1}/_{4}$	3	4	5	6
		Allowable Span (In.)				
8	$10 \, ^{2}/_{3}$	78	89	98	109	113
8	$13 \, ^{1}/_{3}$	84	96	106	114	122
8	16	89	102	113	121	129
12	$10 \, ^{2}/_{3}$	69	79	87	94	100
12	$13 \, ^{1}/_{3}$	74	85	94	101	116
12	16	79	90	100	107	114

Check local codes

Lintels—Concrete, see page 128.

LINTELS—CONCRETE MASONRY

"U" Type Units Filled with Average Strength Concrete, Standard Weight Units

Unit Size	Rods	Safe Superimposed Load/Lin. Ft. 8" Bearing each End										
		Spans										
		3'-4"	4'	4'-8"	5'-4"	6'	6'-8"	7'-4"	8'	8'-8"	9'-4"	10'
7 5/8" × 7 5/8" × 7 5/8"	2-#3	970	655	465	340	255	195					
	1-#3 & 1-#4	980	810	660	490	355	255	190				
	2-#4			675	580	415	300	220				
48 Lbs./Lin. Ft.	1-#4 & 1-#5					480	345	255	195			
	2-#5					495	385	285	220	200		
	1-#5 & 1-#6						430	320	240	210		
	2-#6							340	260			
7 5/8" × 15 5/8" × 7 5/8"	2-#4	2680	2210	1880	1640	1400	1110	900	735	610	510	430
	1-#4 & 1-#5					1420	1270	1140	960	795	675	570
96 Lbs./Lin. Ft.	2-#5								1030	940	830	710
	1-#5 & 1-#6										855	790

Check local codes

LINTELS—CONCRETE

CONCRETE LINTELS WITH WALL LOAD					
	One Piece Lintel		Split Lintel		
Span (Ft.)	Size (In.)	Reinforcing (In.)	Size (In.)	Reinforcing (In.)	End Bearing (In.)
To 7	7 5/8 × 5 3/4	2 - 3/8 Bars	3 5/8 × 5 3/4	1 - 3/8 Bar	8
7 to 8	7 5/8 × 5 3/4	2 - 5/8 Bars	3 5/8 × 5 3/4	1 - 5/8 Bar	8
To 8	7 5/8 × 7 5/8	2 - 3/8 Bars	3 5/8 × 7 5/8	1 - 3/8 Bar	8
8 to 9	7 5/8 × 7 5/8	2 - 1/2 Bars	3 5/8 × 7 5/8	1 - 1/2 Bar	12
9 to 10	7 5/8 × 7 5/8	2 - 5/8 Bars	3 5/8 × 7 5/8	1 - 5/8 Bar	12

CONCRETE LINTELS WITH WALL AND FLOOR LOAD (75 TO 85 POUNDS/SQUARE FOOT, 20 FOOT SPAN

Span (Ft.)	Size (In.)	Bottom Reinforcing (In.)	Top Reinforcing (In.)	Stirrups or Webs	End Bearing (In.)
3	7 5/8 × 7 5/8	2 - 1/2 Bars	—	—	8
4	7 5/8 × 7 5/8	2 - 3/4 Bars	—	3 – 6 ga.	8
5	7 5/8 × 7 5/8	2 - 7/8 Bars	2 - 3/8 Bars	5 – 6 ga.	8
6	7 5/8 × 7 5/8	2 - 7/8 Bars	2 - 1/2 Bars	6 ga.	12
7	7 5/8 × 7 5/8	2 - 1 Bars	2 - 1 Bars	9 ga.	12

LINTELS—FORMED STEEL
3 1/2 × 3 1/2 × 1/8 Inch, 2.9-Pound Stock

Lengths (In.)	Weight (Lbs.)
30	7.3
36	8.7
42	10.2
48	11.6
54	13
60	14.5

LINTELS—PRESSED STEEL
11-Gauge Weights

Lengths	For Block Size (Wt. in Lbs.)				
	4″	6″	8″	10″	12″
30″	4				
36″	5	10	10 1/2	16	17
42″	6	11	12	19	20
48″	7	13	14	21	22
54″	8	14	15	23	24
60″		16	17	26	27
72″			21		

LOAD LIMITS—ALUMINUM SHEET ROOFING

Purlin Spacing	4′	4′-6″	5′	5′-6″	6′
Load (Lbs./Sq. Ft.)	80	60	50	40	35

Check local codes and manufacturer's recommendations

Load Limits—Bar Joists, see page 130.

LOAD LIMITS—CONCRETE-FILLED PIERS

Size	Max. Heights (Ft.)	Max. Load Limit	
		Solid Masonry	Hollow Units
8 × 8	7	15,000	6,200
12 × 12	10	35,000	14,000
16 × 16	13	64,000	25,000
20 × 20	16	100,000	40,000
24 × 24	20	140,000	57,000

Check local codes

LOAD LIMITS—BAR JOISTS
Live & Dead Limits—Pounds/Linear Foot

Span Ft.	Size 8" (3.31)	Size 8" (4.08)	Size 10" (4.02)	Size 10" (4.74)	Size 10" (5.78)	Size 12" (4.94)	Size 12" (5.73)	Size 12" (6.85)	Size 12" (7.92)	Size 14" (6.97)	Size 14" (8.32)	Size 14" (9.43)	Size 16" (8.42)	Size 16" (9.85)	Size 18" (8.80)	Size 18" (9.90)	Size 20" (9.90)
4	800																
5	640																
6	530																
7	400																
8	300																
9	245																
10		350															
11		290															
12		245	290														
13		200	250														
14		175	215														
15		155	185	245		270											
16		135	165	210	260	240											
17			145	190	230	210											
18			130	170	205	190	235										
19				150	185	170	210			290							
20				135	165	155	180	235		260					360	380	390
21						140	175	215		235					340	360	371
22						125	160	195	240	215					325	345	355
23						115	145	180	220	200	260				310	330	340
24							133	165	200	180	240				295	315	325
25										165	220	260	250		270	305	310
26										155	200	245	230		250	290	300
27										145	190	225	210	260	230	280	290
28										135	175	210	200	240	215	265	280
29															200	245	270
30															190	230	250
31															175	215	235
32															165	200	220

Check local codes and manufacturer's specifications

LOAD LIMITS—CONCRETE SLAB FORM (STEEL RIB)
No Added Reinforcement

Total Slab Thickness (In.)	2 1/2		3		3 1/2		4		4 1/2	
Effective Slab Thickness (In.)	1.56		2.06		2.56		3.06		3.56	
Weight of Slab (Lbs./ Sq. Ft.)	32		38		44		50		57	
Gauge of Form										
SPAN	20	18	20	18	20	18	20	18	20	18
					LIVE LOADS (Limits Lbs./Sq. Ft.)					
3' 0"	300	327	491	540	623	792	753	1004	886	1182
3' 6"	212	232	351	387	446	570	540	725	636	854
4' 0"	155	171	260	287	331	426	402	543	474	640
4' 6"	116	128	197	219	253	328	307	419	362	494
5' 0"	88	98	152	170	196	257	239	330	283	390
5' 6"	67	76	119	134	154	205	189	263	224	312
6' 0"	52	58	94	106	123	165	151	213	179	253
6' 6"	39	45	75	85	98	134	121	175	144	208
7' 0"		35	59	68	78	110	97	144	117	171
7' 6"			46	54	63	90	78	119	95	142
8' 0"			36	43	50	73	63	98	76	118
8' 6"					39	60	50	81	61	89
9' 0"					30	49	39	67	49	81
9' 6"						39	30	55	38	67
10' 0"								45	29	55

Check local codes

LOAD LIMITS—CONCRETE SLABS
Pounds/Square Foot

Span (Ft.)	Slab Thickness			
	6″	8″	10″	12″
2	3800	—	—	—
4	900	1750	2800	4200
6	360	700	1200	1800
8	160	360	600	950
10	80	200	350	550
12	—	100	200	350
14	—	—	100	200
16	—	—	—	125
Rods area	.25	.25	.39	.39
Rods spacing	6″	5″	6″	5″
Rods depth	1″	1″	1 1/4″	1 1/2″
Wt. Slab sq. ft.	75	100	125	150

Check local codes

LOAD LIMITS—CORRUGATED STEEL
Safe Load/26-Inch Wide Sheets, Supported at Ends

Gauge	28	27	26	24	22	20	18	16
2 1/2-INCH CORRUGATIONS								
6′ - 0″ Span	88	97	106	141	176	211	282	352
7′ - 0″ Span	75	83	91	121	151	181	242	302
8′ - 0″ Span	60	73	79	106	132	158	211	264
9′ - 0″ Span	59	65	70	94	117	141	188	235
10′ - 0″ Span	53	58	63	85	106	127	169	211
1 1/4-INCH CORRUGATIONS								
6′ - 0″ Span	53	58	63	85	106	127	169	211
7′ - 0″ Span	45	50	55	73	91	109	145	181
8′ - 0″ Span	40	44	47	64	79	95	127	158
9′ - 0″ Span	35	39	42	56	70	85	113	141
10′ - 0″ Span	32	34	38	51	64	76	101	127

Check local codes and manufacturer's recommendations

LOAD LIMITS—METAL LATH

3/4-INCH LATH

Thickness of Slab above mesh (In.)	Wt. of Slab (Lbs./Sq. Ft.)	Wt. with 1/2" P.C. Plaster under Slabs (Lbs.)	Max. Span for Centering Wet Concrete	Safe Loads (Lbs./Sq. Ft.) Span in Feet			
				3'	4'	5'	6'
2	24	30	3' - 3"	325	170	98	59
2	24	30	3' - 7"	438	233	138	87
2 1/2	30	36	2' - 11"	422	222	129	78
2 1/2	30	36	3' - 3"	. . .	302	180	114
3	36	42	2' - 8"	518	273	160	98
3	36	42	2' - 11"	. . .	373	224	142
3 1/2	42	48	2' - 5"	. . .	325	190	117
3 1/2	42	48	2' - 9"	. . .	442	267	170
4	48	54	2' - 3"	. . .	378	222	138
4	48	54	2' - 6"	. . .	514	310	198

3/8-INCH RIB LATH

Thickness of Slab Above Mesh (In.)	Wt. of Concrete (Lbs./Sq. Ft.)	Wt. of Lath (Lbs./Sq. Yd.)	Safe Loads (Lbs./Sq. Ft.) Span in Inches			
			12"	16"	19"	24"
2	24	3.4	950	536	380	238
2	24	4.0	1090	613	433	271
2 1/2	30	3.4	1200	675	479	300
2 1/2	30	4.0	1360	766	544	340
3	36	3.4	1450	815	578	362
3	36	4.0	1650	930	625	412

Check local codes and manufacturer's recommendations

LOAD LIMITS—ROUND STEEL, CONCRETE-FILLED COLUMNS

Safe Loads in Thousands of Pounds/Square Foot

Dia. (In.)	Filled Weight (Lbs./Ft.)	6	7	8	9	10	11	12	13	14	15	16	17	18	19	20	Max. Length (Ft.)
3 1/2	15	37	35	32	29	26	24										11
4	20	49	46	43	40	37	33	30	27								13
4 1/2	24	61	58	55	52	48	45	42	39	35	32						15
5	29	75	72	68	65	61	58	54	51	47	44	40	37				16
5 1/2	36	92	88	84	80	77	73	69	65	62	58	54	50	47	43		18
6 5/8	49	128	124	120	115	111	107	103	99	95	90	86	82	78	74	70	22
7 5/8	64	166	161	156	152	147	143	138	134	129	125	120	115	111	106	102	25
8 5/8	81	211	206	201	196	191	186	181	175	170	165	160	155	150	145	140	29
9 5/8	100	259	253	248	242	237	231	226	221	215	210	204	199	193	188	182	32
10 3/4	123	319	313	307	301	295	289	283	277	271	265	259	253	247	241	236	36
12 3/4	169	421	415	408	402	395	389	382	376	369	363	356	350	343	337	330	43

Check local codes

LOAD LIMITS—SAFETY PLATE

Pounds/Square Foot

Gauge	1'-6"	2'-0"	2'-6"	3'-0"	3'-6"	4'-0"	4'-6"	5'-0"	6'-0"
3/16"	333	188	120	84	61	47	...	...	...
1/4"	593	333	213	148	109	83	66	53	...
5/16"	925	520	333	232	170	130	103	83	58
3/8"	1335	750	480	333	245	188	148	120	84
7/16"	1810	1020	655	453	333	255	204	164	113
1/2"	2370	1330	852	592	435	333	264	213	148
9/16"	3000	1690	1080	750	550	423	333	270	187
5/8"	3700	2080	1330	925	680	520	411	333	232
3/4"	5340	3000	1920	1330	980	750	593	480	333

Check local codes and manufacturer's recommendations

LOAD LIMITS—STEEL BASEMENT COLUMNS

Dia. (In.)	Wall Thickness (In.)	Length	Weight (Lbs.)	Safe Load Capacity (Lbs.) Supporting (with 4″ × 8″ Cap)	
				Steel Beam	Wood Beam
FIXED					
4	1/8	6′	28	22,800	10,000
4	1/8	6′-6″	31	22,500	10,000
4	1/8	7′	34	22,000	10,000
4	1/8	7′-6″	38	21,800	10,000
4	1/8	8′	41	21,200	10,000
ADJUSTABLE					
4	6′-8″ to 7′-1″		38	18,000	10,000
4	7′-2″ to 7′-7″		40	17,700	10,000

Check local codes

LOAD LIMITS—STEEL I BEAMS
Safe Loads on I Beams (Pounds)

Size (In.)	Wt. per Ft. (Lbs.)	6′	7′	8′	9′	10′	11′	12′	13′	14′	15′	16′	17′
4	7.7	2360	1790	1560	1200	1070	970	750	685	515	475	---	---
5	10.0	3840	3280	2540	2245	2015	1525	1445	1125	1035	960	725	675
6	12.5	5770	4940	4305	3390	3035	2745	2185	2005	1570	1455	1350	1035
7	15.3	8235	7045	6150	5455	4340	3935	3590	2875	2655	2465	1950	1820
8	18.4	12585	9690	8460	7500	6735	5415	4950	4550	4205	3410	3180	2530
9	21.8	16735	14325	11215	9970	8955	8120	6580	6055	5605	4540	4235	3965
10	25.4	21625	18510	14570	12890	11600	10520	8535	7855	7265	6755	5500	5150
12	31.8	31855	25240	22075	17590	17110	15525	14200	11600	10740	9995	9340	7630

Size (In.)	Wt. per Ft. (Lbs.)	18′	19′	20′	21′	22′	23′	24′	25′	26′	27′	28′	29′
4	7.7	---	---	---	---	---	---	---	---	---	---	---	---
5	10.0	---	---	---	---	---	---	---	---	---	---	---	---
6	12.5	965	905	---	---	---	---	---	---	---	---	---	---
7	15.3	1705	1310	1230	1155	---	---	---	---	---	---	---	---
8	18.4	2370	2230	2100	1620	1530	1445	---	---	---	---	---	---
9	21.8	3725	2975	2810	2655	2515	1955	1845	1750	---	---	---	---
10	25.4	4840	4560	3655	3455	3275	3110	2410	2290	2180	2075	---	2840
12	31.8	7165	6760	6390	5140	4875	4630	4410	4200	3270	3115	2975	2840

Check local codes

LOAD LIMITS—STEEL COLUMNS

Weights, Load Limits, Areas

3 IN. O.D. 11 GA. STEEL AREA 1.086 SQ. IN.

Lengths	Weights (Lbs.)	Rec. Safe Load (Lbs./Sq. Ft.)
6'	22	15,900
6'-6"	24	15,400
7'	26	15,000
7'-6"	28	14,300
8'	30	13,700

4 IN. O.D. 11 GA. STEEL AREA 1.46 SQ. IN.

Lengths	Weight (Lbs.)	Rec. Safe Load (Lbs./Sq. Ft.)
6'	30	23,000
6'-6"	32	22,600
7'	35	22,200
7'-6"	37	22,000
8'	40	21,500

LOAD LIMITS—STEEL ROOF DECK

Safe Loads, Pounds/Square Foot

SINGLE SPAN

Rib Depth	Gauge	Wt. per Sq. Ft.	4'0"	4'6"	5'0"	5'6"	6'0"	6'6"	7'0"	7'6"	8'0"	8'6"	9'0"	9'6"	10'0"
1 1/2	18	2.89	105	83	67	55	47	40	34	30					
	20	2.18	79	63	51	42	35	30							
	22	1.81	69	54	44	36	30	26							
1 3/4	18	3.07	138	109	88	73	61	52	45	39	35	31			
	20	2.30	104	82	67	55	46	39	34	30					
	22	1.92	86	68	55	45	38	33	28						
2	18	3.26				93	78	66	57	50	44	39	35	31	28
	20	2.45				70	58	50	43	37	33	29	26		
2 1/2	18	3.57				138	116	99	85	74	65	58	52	46	42
	20	2.68				103	87	74	64	55	49	43	38	35	31

CONTINUOUS SPAN

Rib Depth	Gauge	Wt. per Sq. Ft.	4'0"	4'6"	5'0"	5'6"	6'0"	6'6"	7'0"	7'5"
1 1/2	18	2.89	131	103	84	69	58	50	43	37
	20	2.18	99	78	63	52	44	38	32	28
	22	1.81	88	68	55	45	38	32	28	
1 3/4	18	3.07	173	136	110	91	77	65	56	49
	20	2.30	130	103	83	69	58	49	42	37
	22	1.92	107	85	69	57	48	41	35	31
2 1/2	18	3.57				173	145	124	107	93
	20	2.68				129	108	92	79	69

Check local codes

LOADS LIMITS—TIMBER COLUMNS
Allowable Loads, Pounds/Square Inch

Species of Timber	American Standard Grade	Ratio of Length to Least Dimension										
		10 and Less	12	14	16	18	20	25	30	35	40	50
Ash, Commercial White	(Select	1100	1076	1055	1023	978	913	658	457	336	257	164
	(No. 1	880	868	857	840	818	784	647				
Cedar, Western Red; Fir, Balsam	(Select	700	686	674	656	629	592	438	304	224	171	110
	(No. 1	560	553	547	538	524	505	425				
Cedar, Northern and Southern White	(Select	550	540	530	516	496	468	351	244	179	137	88
	(No. 1	440	435	430	423	412	398	338				
Chestnut; Pine, Northern White, Idaho White, Sugar, Calif. White, 257 and Pondosa	(Select	750	733	718	695	663	617	438	304	224	171	110
	(No. 1	600	591	583	572	556	532	434				
Cypress, Southern; Larch, Western	(Select	1100	1063	1030	981	909	810	526	365	268	206	132
	(No. 1	880	861	843	818	781	729					

Species	Grade											
Douglas Fir (Coast Region); Pine, Southern Yellow; Beech; Birch, Yellow and Sweet; Maple, Sugar	(Dense) (Select)	1285	1251	1222	1176	1112	1022	702	487	358	274	175
	(Select)	1175	1149	1127	1093	1045	975	702				
	(No. 1)	880	870	861	847	826	796	675				
Douglas Fir (Rocky Mtn. Region) Spruce, Red, White, Sitka; Norway Pine; Alaska Cedar, Elm, Slippery and White; Sycamore; Gum, Red and Black; Tupelo	(Select)	800	786	774	753	726	688	526	365	268	206	132
	(No. 1)	640	632	627	617	602	582	500				
Hemlock, West Coast	(Select)	900	885	872	852	823	783	614	426	313	240	153
	(No. 1)	720	712	706	696	680	660	573				
Hemlock, Eastern; Fir, Commercial White	(Select)	700	689	678	664	641	611	482	335	246	188	121
	(No. 1)	560	554	549	542	530	515	449				
Oak, White and Red	(Select)	1000	982	967	943	908	860	658	457	336	257	164
	(No. 1)	800	790	783	771	753	728	625				
Redwood	(Select)	1000	972	947	910	856	781	526	365	268	206	132
	(No. 1)	800	786	773	754	726	688					
Spruce, Englemann	(Select)	600	586	574	556	530	494	351	224	179	137	88
	(No. 1)	480	473	466	457	444	426	347				
Tamarack	(Select)	1000	976	955	923	877	817	570	396	291	223	142
	(No. 1)	800	788	777	761	737	706	566				

Check local codes and current grades

LOAD LIMITS—WIRE MESH (PAPER BACKED)
Pounds/Square Foot (3- x -4-Inch Mesh, 12 Gauge)

Span (In.)	Concrete 2" Slab (Wt: 25 Lb./Sq. Ft.)	Concrete 2 1/2" Slab (Wt: 31 Lb./Sq. Ft.)	Concrete 3" Slab (Wt: 38 Lb./Sq. Ft.)	Concrete 3 1/2" Slab (Wt: 44 Lb./Sq. Ft.)
18	436	586	736	886
20	353	475	597	718
22	292	392	494	594
24	245	330	415	499
26	209	281	353	426
28	180	242	305	366
30	157	211	266	319
32	139	186	234	281
34	123	165	207	248
36	109	146	184	222

Check manufacturer's specifications

LOAD LIMITS—WOOD BEAMS
Safe Loads on Pine Beams Pounds, Load at Center

Span in Feet	7	8	9	10	11	12	13	14	15	16	17	18	19	20
Size Beam (In.)														
2 × 6	400	350	305	275	255	225	205	190	175	160	150	135	125	115
2 × 8	740	645	570	510	460	420	385	355	330	300	285	265	245	230
2 × 10	1185	1030	915	815	740	675	620	570	525	490	460	430	405	375
2 × 12	1370	1135	1010	895	815	740	680	630	575	540	500	475	445	400
4 × 4	365	320	280	250	220	200	185	165	150	140	125	115	105	95
4 × 6	895	775	685	610	555	500	460	420	385	360	330	310	285	270
4 × 8	1645	1440	1270	1135	1030	935	855	785	730	675	625	585	550	515
6 × 6	1385	1205	1065	945	850	775	710	650	600	550	510	475	445	415
6 × 8	2560	2225	1970	1760	1590	1450	1325	1220	1115	1045	975	910	850	795
6 × 10	4085	3560	3155	2825	2555	2325	2135	1965	1825	1700	1580	1480	1385	1305
8 × 8	3465	3015	2675	2385	2160	1960	1800	1655	1530	1420	1315	1230	1150	1080
8 × 10	5540	4830	4275	3830	3460	3155	2890	2670	2475	2300	2150	2010	1885	1770
8 × 12	8095	7065	6255	5610	5080	4635	4255	3930	3640	3390	3170	2970	2790	2630

Check local codes

LOAD LIMITS—WOOD COLUMNS
In Thousands of Pounds

Length (Feet)	W. PINE SQUARES							OAK SQUARES							FIR-HEMLOCK SQUARES						
	4"	6"	8"	10"	12"	14"	16"	4"	6"	8"	10"	12"	14"	16"	4"	6"	8"	10"	12"	14"	16"
5	12							16							14						
6	11							15							13						
7	10							13							12						
8	9½	26						12	34						11	32					
9	9	25						11	33						10	30					
10	8	24	48					10	31	62					9	29	58				
11		23	46						30	60						27	56				
12		22	45						28	58						26	54				
14		19	42	72					25	54	94					23	50	86			
16			38	68	105					50	88	137					46	82	127		
18			35	64	101	146				46	83	131	189				42	77	121	174	
20			32	60	96	140	192			42	78	125	182	250			38	72	115	168	230

Check local codes

LOAD LIMITS—WOOD JOISTS
Limit/Square Foot, 1200-Pound Fiber Stress

Size	Span						
	8	10	12	14	16	18	20
6"	90	70	50	35	28	—	—
8"	160	125	80	60	48	38	30
10"	—	200	130	100	78	63	50
12"	—	—	200	150	118	92	70

LOAD LIMITS—WOOD STUDS
Safe Loads/Linear Foot of Wall

Size	Spacing	Height	Safe Load	Weight	(Frame Only) Bd. Ft.
2 × 4	12"	8	3700	16	6 2/3
	12	10	3200	20	8
	12	12	2600	23	9 1/3
	16	8	2800	13	5 1/3
	16	10	2400	15	6 1/3
	16	12	2000	18	7 1/3
2 × 6	12"	8	5800	25	
	12	10	4900	31	
	12	12	4100	35	
	16	8	4300	20	
	16	10	3700	24	
	16	12	3000	28	

Check local codes

LOAD LIMITS—WOOD STUDS, JOISTS, RAFTERS
Safe Loads/Square Foot

2 × 8's and 2 × 6's — Spacing on Centers

Span	2×8 12"	2×8 14"	2×8 16"	2×8 18"	2×8 20"	2×6 12"	2×6 14"	2×6 16"	2×6 18"	2×6 20"
6 ft.	401	343	300	264	239	227	195	170	151	136
7 ft.	297	254	222	197	177	165	142	123	109	98
8 ft.	226	194	169	150	134	109	94	82	72	65
9 ft.	179	153	134	118	106	78	65	57	50	45
10 ft.	132	113	98	87	81	55	47	41	35	32
11 ft.	98	84	73	64	58	41	34	29	26	24
12 ft.	75	64	56	49	44	31	26	22	X	X
13 ft.	59	50	44	38	34	23	20	X	X	X
14 ft.	46	39	34	30	27	X	X	X	X	X
15 ft.	37	31	27	24	21	X	X	X	X	X
16 ft.	30	25	22	X	X	X	X	X	X	X

2 × 12's and 2 × 10's — Spacing on Centers

Span	2×12 12"	2×12 14"	2×12 16"	2×12 18"	2×12 20"	2×10 12"	2×10 14"	2×10 16"	2×10 18"	2×10 20"
10 ft.	341	292	260	231	208	232	195	170	138	135
11 ft.	281	240	209	185	166	191	163	142	126	113
12 ft.	235	200	180	155	139	156	133	116	103	92
13 ft.	190	176	148	132	113	121	103	90	79	71
14 ft.	172	147	128	113	101	96	82	71	63	56
15 ft.	140	124	104	92	82	78	66	57	51	45
16 ft.	115	98	85	75	67	63	53	46	41	36
17 ft.	94	80	70	61	55	52	44	38	33	30
18 ft.	79	68	58	51	45	43	36	31	27	24
19 ft.	66	56	48	42	38	36	30	26	23	20
20 ft.	56	47	41	36	32	30	25	22	X	X

Check local codes

LOADS—DEAD

Weights of Various Materials/Square Foot for Use in Determining Dead Loads on Floors and Roofs

Material	Weight (Lb./Sq. Ft.)	Material	Weight (Lb./Sq. Ft.)
Boards, fiber insulating, 1″	1.50	Roofing, built-up	2 to 7
Boards, fiber insulating, 3/4″	1.10	Roofing, composition shingles	3.
Boards, fiber insulating, 1/2″	0.80		
Ceiling, wood, 3/4″	2.50	Roofing, wood shingles	2.
Ceiling, wood, 5/8″	1.80	Roofing, tile or slate	5 to 20
Ceiling, wood, 1/2″	1.40	Roof slabs of precast concrete	15.
Ceiling, wood, 3/8″	1.10		
Ceiling, gypsum lath & plaster (3/8″ plus 1/2″ thick)	5.50	Roofs, asphalt, felt & gravel (3-5 ply built-up)	5-6 1/2
Ceiling, lath & 3/4″ plaster	8.	Roofs, asphalt, felt & slag (3-5 ply built-up)	4 1/2 - 5 1/2
Ceiling, suspended metal lath and plaster	10.	Roofs, composition 3-ply (ready roofing)	1.
Ceiling, gypsum dry-wall board (1/2″ thick)	2.	Roofs, concrete, cinder (per inch thickness)	9.
Copper, sheet	1.	Roofs, concrete, nailing (per inch thickness)	8.
Floor finish of terrazzo, tile per in. thickness including base	12.	Roofs, corrugated aluminum (.024″ thick)	1/2
Hardwood finish	4.	Roofs, corrugated iron-steel (20-18 gauge)	2-3
Iron, corrugated	1 to 4	Roofs, sheathing boards (1″ WP. Spr. Hmlk.)	2 1/2-3
Iron, flat seam, galvanized	1 to 3		
Joists - 2 × 8 @ 20 in.	2.	Roof, shingles, asphalt, slate-covered	2.
2 × 8 @ 14 in.	3.	Roof, shingles, wood	2-3
2 × 10 @ 18 in.	3.	Roof, slate (3/16″ - 3/8″ thick)	7-14
2 × 10 @ 12 in.	4.	Roof, tile, clay	10-15
2 × 12 @ 16 in.	4.	Sheathing, wood, 3/4″	2.50
2 × 12 @ 12 in.	5.	Sheathing, wood, 1 5/8″	5.40
Lead, sheet	4 to 8	Shingles, asphalt	2-3
Plaster on masonry, one side	5.	Shingles, wood	2.50
		Slate	10.
Plaster, wood lath, 3/4″ grounds	5.	Sheathing	3.
		Stucco	10.
Plaster, fiberboard lath, 1″ grounds	5.	Studs, plates & bridging 2 × 4 @ 12 in. to 16 in.	2.
Plaster, gypsum board lath, 7/8″ grounds	6.	2 × 6 @ 12 in. to 16 in.	3.
Plaster, metal lath, 3/4″ grounds	6.	Suspended ceiling - metal lath and plaster	10.
Plywood, 1/4″	0.70	Tile, plain	9-12
Plywood, 5/16″	1.	Tin, painted	1.
Plywood, 3/8″	1.10	Zinc, sheet	1-2
Roofing, heavy rool	1.		
Roofing, tar and gravel	6.		

LOADS—LIVE

Recommended Uniform Roof Live Loads for APA RATED SHEATHING and APA RATED STURD-I-FLOOR With Long Dimension Perpendicular to Supports[c]

Panel Span Rating	Panel Thickness (In.)	Maximum Span (In.) With Edge Support[a]	Maximum Span (In.) Without Edge Support	Allowable Live Loads (psf)[d] — Spacing of Supports Center-to-Center (In.) 12	16	20	24	32	40	48	60
APA RATED SHEATHING											
12/0	5/16	12	12	30							
16/0	5/16, 3/8	16	16	70	30						
20/0	5/16, 3/8	20	20	120	50	30					
24/0	3/8, 7/16, 15/32, 1/2	24	20(b)	190	100	60	30				
24/16	7/16, 1/2	24	24	190	100	65	40	30			
32/16	15/32, 1/2, 5/8	32	28	325	180	120	70	40	30		
40/20	19/32, 5/8, 3/4, 7/8	40	32	—	305	205	130	70	60	30	
48/24	23/32, 3/4, 7/8	48	36	—	—	410	280	175	95	45	35
APA RATED STURD-I-FLOOR											
16 oc	19/32, 5/8, 21/32	24	24	185	100	65	40				
20 oc	19/32, 5/8, 3/4	32	32	270	150	100	60	30			
24 oc	11/16, 23/32, 3/4	40	36	—	240	160	100	50	30		
32 oc	7/8, 1	48	40	—	430	295	185	100	60	40	
48 oc	1 3/32, 1 1/8	60	48	—	—	—	290	160	100	65	40

(a) Tongue-and-groove edges, panel edge clips (one between each support, except two between supports 48 inches on center), lumber blocking, or other.

(b) 24 inches for 15/32-inch and 1/2-inch panels.

(c) When roofing is to be guaranteed by a performance bond, check with roofing manufacturer for panel type, minimum thickness span and edge support requirements.

(d) 10 psf dead load assumed.

LOADS—LIVE (HUMAN OCCUPANCY)

	Weight (Lbs./Sq. Ft.)
HUMAN OCCUPANCY	
Dwellings	40
Hospital Rooms and Wards	40
Hotel Guest Rooms and Lobbies	40
Tenements	40
Office Buildings	50
School Rooms	50
Corridors in hospitals, hotels, schools, etc.	100
Assembly Rooms with fixed seats	50
Grandstands	100
Theatre stages	100
Gymnasiums	100
Stairways and Fire Escapes	100
INDUSTRIAL OR COMMERCIAL OCCUPANCY	
Storage Purposes (General)	250
Storage Purposes (Special)	125
Manufacturing (Light)	75
Printing Plants	100
Wholesale Stores (Light merchandise)	100
Retail Storerooms (Light merchandise)	75
Stables	75
Garages, all types of vehicles	100
Garages, Passenger cars only	50
Sidewalks	250
Roofs, up to slope of 4 in. per ft.	20
Roofs, slope of 4 in. to 12 in. per ft.	16
Roofs, slope of over 12 in. per ft. wind force of 20 pounds per sq. ft. acting normal to surface	12

Check local codes

LOADS—LIVE (WAREHOUSE FLOORS)
Average Piling 5 to 6 Feet High

Materials	Approx. Wt. (Lbs.)/ Cu. Ft.	Wt. (Lbs.)/Sq. Ft. 5′ to 6′ Pile P.S.F.
Asbestos Sheets	50 – 60	250 – 300
Common Brick	40 – 50	200 – 300
Fire Brick	70 – 80	350 – 500
Cement (Portland)	90 – 100	450 – 600
Cement (Masonry)	70 – 80	350 – 450
Plasters (Cement)	60 – 70	300 – 400
Lime	45 – 55	250 – 350
Lumber	40 – 60	200 – 350

LOADS—SAFE BEARING

Load	Lbs./ Sq. In.
BRICK AND STONE MASONRY BRICKWORK	
Bricks, hard, laid in lime mortar	100
Hard, laid in Portland Cement mortar	200
Hard, laid in Rosedale Cement mortar	150
MASONRY	
Granite, capstone	700
Squared stonework	350
Sandstone, capstone	350
Squared stonework	175
Rubble stonework, laid in lime mortar	80
Rubble stonework, laid in cement mortar	150
Limestone, capstone	500
Squared stonework	250
Rubble, laid in lime mortar	80
Rubble, laid in cement mortar	150
Concrete, 1 Portland, 2 sand, 5 broken stone	150

	Tons/ Sq. Ft.
FOUNDATION	
Rock, hardest in native bed	100
Equal to best ashlar masonry	25-40
Equal to best brick	15-20
Clay, dry in thick beds	4-6
Moderately dry, in thick beds	2-4
Soft	1-2
Gravel and coarse sand, well cemented	8-10
Sand, compact and well cemented	4-6
Clean, dry	2-4
Quicksand, alluvial soils. etc.	5-1

Check local codes

LUMBER—AVERAGE WEIGHTS (MBF PINE)

Type Lumber	Short Leaf
DRY	
Flooring, $^{25}/_{32} \times 2 \, ^{3}/_{8}$)	(1800
Flooring, $^{25}/_{32} \times 3 \, ^{1}/_{4}$)	(1900
Flooring, $^{25}/_{32} \times 5 \, ^{1}/_{4}$) For Hollow Back Flooring,	(2100
Ceiling, $^{5}/_{16}$ ....) Ceiling and Drop Siding,	(900
Ceiling, $^{7}/_{16}$ ....) Deduct 100 lbs.	(1100
Ceiling, $^{9}/_{16}$ ....)	(1400
Ceiling, $^{11}/_{16}$ ....)	(1700
Partition, $^{3}/_{4}$	1800
Siding, from inch stock.............................	1000
Siding, from 1 $^{1}/_{4}$ inch stock.............................	1250
Drop Siding to $^{3}/_{4}$	—
Drop Siding, $^{3}/_{4}$ and Moulded Casing	1800
Moulded Base...............................	2000
Finish, inch, S1S or S2S to $^{25}/_{32}$	2400
Finish, 1 $^{1}/_{4}$, 1 $^{1}/_{2}$ and 2 inch, rough...................	3200
Industrial Standard 1 inch, S1S or S2S $^{13}/_{16}$	2500
Industrial Standard 2 inch, S1S or S2S 1 $^{3}/_{4}$	2900
SHIPPING WTS. DRY	
1 $\times$ 4, S2s and C.M., $^{25}/_{32}$	2100
1 $\times$ 6, S1S and C.M., $^{25}/_{32}$	2200
Shiplap, and D. & M. $^{25}/_{32}$	2300
Grooved Roofing...............................	2300
Common Boards & Fencing, 1 $\times$ 4, 6, 8 & 10 in. S1S or S2S to $^{25}/_{32}$	2400
Common Boards, 1 $\times$ 12, S1S or S2S to $^{25}/_{32}$	2500
Common Boards and Fencing, 1 $\times$ 4, 6, 8, or 10 inch, rough	3300
Common Boards, 1 $\times$ 12, rough	3400
2 $\times$ 4, 2 $\times$ 6 and 2 $\times$ 8, S1S1E to 1 $^{5}/_{8}$...............	2500
2 $\times$ 4, 2 $\times$ 6 and 2 $\times$ 8, rough...................	3300
2 $\times$ 10 and 2 $\times$ 12, S1S1E to 1 $^{5}/_{8}$	2600
GREEN	
2 $\times$ 14 and 3 $\times$ 12, S1S1E	3500
2 $\times$ 14 and 3 $\times$ 12, rough	4200
3 $\times$ 4 and 6 $\times$ 6, S1S1E	3500
4 $\times$ 4 and 6 $\times$ 8, rough...................	4200

LUMBER—BOARD FEET CALCULATIONS

Size of Piece (In.)	Length in Feet					
	10	12	14	16	18	20
1 × 2	1 2/3	2	2 1/3	2 2/3	3	3 1/3
1 × 3	2 1/2	3	3 1/2	4	4 1/2	5
1 × 4	3 1/3	4	4 2/3	5 1/3	6	6 2/3
1 × 5	4 1/6	5	5 5/6	6 2/3	7 1/2	8 1/3
1 × 6	5	6	7	8	9	10
1 × 8	6 2/3	8	9 1/3	10 2/3	12	13 1/3
1 × 10	8 1/3	10	11 2/3	13 1/3	15	16 2/3
1 × 12	10	12	14	16	18	20
1 × 14	11 2/3	14	16 1/3	18 2/3	21	23 1/3
1 × 16	13 1/3	16	18 2/3	21 1/3	24	26 2/3
2 × 4	6 2/3	8	9 1/3	10 2/3	12	13 1/3
2 × 6	10	12	14	16	18	20
2 × 8	13 1/3	16	18 2/3	21 1/3	24	26 2/3
2 × 10	16 2/3	20	23 1/3	26 2/3	30	33 1/3
2 × 12	20	24	28	32	36	40
2 × 14	23 1/3	28	32 2/3	37 1/3	42	46 2/3
2 × 16	26 2/3	32	37 1/2	42 2/3	48	53 1/3
3 × 6	15	18	21	24	27	30
3 × 8	20	24	28	32	36	40
3 × 10	25	30	35	40	45	50
3 × 12	30	36	42	48	54	60
3 × 14	35	42	49	56	63	70
3 × 16	40	48	56	64	72	80
4 × 4	13 1/3	16	18 2/3	21 1/3	24	26 2/3
4 × 6	20	24	28	32	36	40
4 × 8	26 2/3	32	37 1/3	42 2/3	48	53 1/3
4 × 10	33 1/3	40	46 2/3	53 1/3	60	66 2/3
4 × 12	40	48	56	64	72	80
4 × 14	46 2/3	56	65 1/3	74 2/3	84	93 1/3
6 × 6	30	36	42	48	54	60
6 × 8	40	48	56	64	72	80
6 × 10	50	60	70	80	90	100
6 × 12	60	72	84	96	108	120

LUMBER—DETERMINING QUANTITIES

STANDARD FLOORING: Multiply area (in square feet) by conversion factors given —

Nominal Thickness and Width	Net Thickness and Width	Conversion Factor	Wt. M ft. Oak-Maple
1 × 3	25/32 × 2 3/8″	1.301	2000#
1 × 4	25/32 × 3 1/4″	1.268	2250#
1 × 6	25/32 × 5 3/16″	1.192	2300#
1 1/4 × 3	1 1/16 × 2 3/8″	1.626	2300#
1 1/4 × 4	1 1/16 × 3 1/4″	1.585	2300#
1 1/4 × 6	1 1/16 × 5 3/16″	1.489	2350#

BEVEL SIDING: Compute area to be covered in (square feet) and deduct for openings in the wall. Multiply by conversion factors.

1/2 × 4″ bevel siding, figuring lap cover of 3/4″ 1.498
1/2 × 6″ bevel siding, figuring lap cover of 1″ 1.373
1/2 × 8″ bevel siding, figuring lap cover of 1 1/2″ 1.433

1 × 6 DROP SIDING:

Pattern 105 lays with a net face width of 5 1/16″ 1.220
Pattern 106 lays with a net face width of 5 3/16″ 1.192

INCH BOARDS & SHIPLAP:

1 × 6″ boards, S4S standard 3/4 × 5 5/8″ 1.099
1 × 8″ boards, S4S standard 3/4 × 7 1/2″ 1.099
1 × 6″ shiplap, finished 3/4 × 5 1/8″ face width. 1.206
1 × 8″ shiplap, finished 3/4 × 6 7/8″ face width. 1.156

LUMBER—FINISHED SIZES REQUIRED

Size, (In.)	Fin. Width, (In.)	Add for Waste (%)	Req. Multiply Area by	Lbr. Req. 100 Sq. Ft.
1 × 3	2 1/2	25	1.25	125
1 × 4	3 1/2	20	1.20	120
1 × 6	5 1/2	14	1.14	114
1 × 8	7 1/4	12	1.12	112
1 × 10	9 1/4	10	1.10	110
1 × 12	11 1/4	9 1/2	1.095	109 1/2
2 × 4	3 1/2	20	2.40	240
2 × 6	5 1/2	14	2.28	228
2 × 8	7 1/4	12	2.25	225
2 × 10	9 1/4	10	2.20	220
2 × 12	11 1/4	9 1/2	2.19	219
3 × 6	5 1/2	14	3.43	343
3 × 8	7 1/4	12	3.375	337 1/2
3 × 10	9 1/4	10	3.30	330
3 × 12	11 1/4	9 1/2	3.29	329

LUMBER—PRESERVATIVE PRESSURE-TREATED
Preservative Retention Requirements (Pounds Chemical/ Cubic Feet)

Application	Retention
Above Ground	0.25
Ground Contact	0.40
Fresh Water Exposure	0.40
Salt Water Splash	0.40
Permanent Wood Foundation	0.60
Brackish Water Immersion	1.00
Salt Water Immersion	2.50

LUMBER—STANDARD SIZES (DRY-DRESSED S4S)

Type	Thickness	Width	Actual Size Thickness	Width
Dimension	2 in.	4 in.	1 1/2	3 1/2
	2 in.	6 in.	1 1/2	5 1/2
	2 in.	8 in.	1 1/2	7 1/4
	2 in.	10 in.	1 1/2	9 1/4
	2 in.	12 in.	1 1/2	11 1/4
Timbers	4 in.	6 in.	3 1/2	5 1/2
	4 in.	8 in.	3 1/2	7 1/2
	4 in.	10 in.	3 1/2	9 1/2
	6 in.	6 in.	5 1/2	5 1/2
	6 in.	8 in.	5 1/2	7 1/2
	6 in.	10 in.	5 1/2	9 1/2
	8 in.	8 in.	7 1/2	7 1/2
	8 in.	10 in.	7 1/2	9 1/2
Common Boards	1 in.	4 in.	3/4	3 1/2
	1 in.	6 in.	3/4	5 1/2
	1 in.	8 in.	3/4	7 1/4
	1 in.	10 in.	3/4	9 1/4
	1 in.	12 in.	3/4	11 1/4
Shiplap Boards	1 in.	4 in.	3/4	3 1/8
	1 in.	6 in.	3/4	5 1/8
	1 in.	8 in.	3/4	6 7/8
	1 in.	10 in.	3/4	8 7/8
	1 in.	12 in.	3/4	10 7/8
Tongued and Grooved Boards	1 in.	4 in.	3/4	3 1/8
	1 in.	6 in.	3/4	5 1/8
	1 in.	8 in.	3/4	6 7/8
	1 in.	10 in.	3/4	8 7/8
	1 in.	12 in.	3/4	10 7/8

LUMBER—TONGUE & GROOVE, SHIPLAP
(SIZES REQUIRED)

Size (In.)	Width (In.)	Add for Waste (%)	Req. Multiply Area by	Req. 100 Sq. Ft. Surface
1 × 2	1 3/8	50	1.50	150
1 × 2 3/4	2	42 1/2	1.425	142 1/2
1 × 3	2 1/4	38 1/3	1.383	138
1 × 4	3 1/4	28	1.28	128
1 × 6	5 1/4	20	1.20	120
1 × 8	7 1/4	16	1.15	115
1 1/4 × 3	2 1/4	38 1/3	1.73	173
1 1/4 × 4	3 1/4	28	1.60	160
1 1/4 × 6	5 1/4	20	1.50	150
1 1/2 × 3	2 1/4	38 1/3	2.08	208
1 1/2 × 4	3 1/4	28	1.92	192
1 1/2 × 6	5 1/4	20	1.80	180
2 × 4	3 1/4	28	2.60	260
2 × 6	5 1/4	20	2.40	240
2 × 8	7 1/4	16	2.32	232
2 × 10	9 1/4	13	2.25	225
3 × 12	11 1/4	12	2.24	224

LUMBER SIDING—DETERMINING QUANTITIES

1-INCH BOARDS		
Lumber Size (In.)	Finished Size (In.)	Percent to Add
6	5 1/2	14
8	7 1/4	12
10	9 1/4	10
12	11 1/4	9 1/2

1-INCH SHIPLAP		
Lumber Size (In.)	Face Width (In.)	Percent to Add
8	7 1/4	16
10	9 1/4	13
12	11 1/4	12 1/2

D & M OR CEILING

4	3 1/4	28
6	5 1/4	20

LAP SIDING

Siding Width	Exposure	Percent to Add
4	2 3/4	50
6	4 3/4	31
8	6 3/4	25
10	8 3/4	20

M

MASONRY—BRICK
Number/1000 Square Feet

Width of Joint (In.)	Thickness of Wall (In.)			
	4	8	12	16
1/4	6,980	14,000	20,900	27,000
3/8	6,550	13,100	19,600	26,200
1/2	6,160	12,300	18,500	24,600
5/8	5,810	11,600	17,400	23,200
3/4	5,490	11,000	16,500	21,900

MASONRY—BRICK
Square Feet/1000 Brick

Width of Joint (In.)	Thickness of Wall (In.)			
	4	8	12	16
1/4	143	72	48	36
3/8	153	76	51	38
1/2	162	81	54	41
5/8	172	86	57	43
3/4	182	91	61	46

MASONRY—BRICK (JUMBO)

Sizes (In.)	No./ Sq. Ft. Wall	Mortar/1000 Brick (1/2-In. Jts.)
7 1/2 × 3 1/2 × 11 1/2	3	17 1/2 cu. ft.
3 1/2 × 3 1/2 × 11 1/2	3	15 1/2 cu. ft.
5 1/2 × 3 1/2 × 11 1/2	3	18 cu. ft.
2 1/2 × 3 1/2 × 11 1/2	3	11 1/2 cu. ft.

MASONRY—BRICK (MORTAR REQUIREMENTS)
Quantities/1000 Brick

Width of Joint (In.)	Thickness of Wall (In.)			
	4		8,12,16 & Larger	
	Cu. Ft.	Cu. Yd.	Cu. Ft.	Cu. Yd.
1/4	5.7	0.21	8.7	0.32
3/8	8.7	0.32	11.8	0.44
1/2	11.7	0.43	15.0	0.56
5/8	14.8	0.55	18.3	0.68
3/4	17.9	0.66	21.7	0.80

MASONRY—BRICK (SCR)
Quantities/100 Square Feet

- 100 Sq. Ft. Wall—1/2″ Jts.
- 450 Brick SCR
- 11 Cu. Ft. Mortar

MASONRY—BRICK (SCR)
Quantity/100 Square Feet of Wall, 1/2-Inch Joints

- Bricks—450
- Mortar—16 cu. ft.
- 3/strip, 16-in. O.C.

MASONRY—BRICK (TYPICAL SIZES & WEIGHTS)

Type	Size	Weight (Lbs.)		Tons per 1000
		Each	Per 1000	
Modular Conc.	2 1/4″ × 3 5/8″ × 7 5/8″	5	5,000	2.5
Common Reg.	2 1/4″ × 3 3/4″ × 8″	5 1/2	5,500	2.75
Common Hard	2 1/4″ × 3 3/4″ × 8″	6 1/2	6,500	3.25
Common Soft	2 1/4″ × 4″ × 8 1/4″	4 1/4	4,250	2.12
Paving Brick	2 1/2″ × 4″ × 8 1/2″	6 3/4	6,750	3.27
Fire Brick Std.	2 1/2″ × 4 1/2″ × 9″	7	7,000	3.5
Fire Brick Split	1 1/4″ × 4 1/2″ × 9″	3 1/2	3,500	1.75
Face Brick	2 1/8″ × 3 5/8″ × 8″	5	5,000	2.5
Normans	2 1/4″ × 3 1/2″ × 11 1/2″	6 1/2	6,500	3.25
Romans	1 3/4″ × 3 1/2″ × 11 1/2″	5	5,000	2.5
S.C.R.	2 1/6″ × 5 1/2″ × 11 1/2″	9	9,000	4.5
	(2 1/2″ × 4″ × 6″)			
Slump	(4″ × 4″ × 12″)			
	(4″ × 4″ × 16″)			

Sizes and weights of brick may vary in different localities.

MASONRY—BRICK BONDS
Brick/100 Square Feet Solid Brick Wall, 1/2-Inch Mortar Joints

Type Bond	Face Brick	8″ Wall	12″ Wall	16″ Wall
Common (Header every 7th course)	700	525	1145	2290
Running	620	620	1230	2460
English (Full headers every 6th course)	720	515	1130	2260
Flemish (Full headers every 5th course)	660	575	1190	2380
Double Headers (Alternating with stretchers every 5th course)	680	555	1170	2340

Add for   1/8″ joints 21%
        1/4″ joints 14%
        3/8″ joints 7%

Subtract for   5/8″ joints 5%
        3/4″ joints 10%
        7/8″ joints 15%
        1″ joints 20%

MASONRY—BRICK PIERS (SIZES, QUANTITY)

1/2-Inch Joints

Feet High	8" × 8"			8" × 12"			12" × 12"			12" × 16"			16" × 16"		
	Face	Common	Total	Face	Common	Total	Face	Common	Total	Face	Common	Total	Face	Common	Total
1	9	0	9	14	0	14	18	2	20	24	4	28	28	8	36
2	18	0	18	27	0	27	36	5	41	43	9	54	54	18	72
3	26	0	26	39	0	39	52	7	59	65	13	78	78	26	104
4	35	0	35	53	0	53	70	9	79	88	18	106	104	36	140
5	44	0	44	66	0	66	88	11	99	110	22	132	132	44	176
6	52	0	52	79	0	79	104	13	117	132	26	158	156	52	208
7	61	0	61	92	0	92	122	15	137	153	31	184	182	62	244
8	70	0	70	105	0	105	140	18	158	175	35	210	210	70	280
9	79	0	79	119	0	119	157	20	177	199	39	238	238	78	316
10	88	0	88	132	0	132	176	22	198	220	44	264	264	88	352

MASONRY—BRICK WALLS
Concrete Masonry Backup

Concrete Masonry Backup Courses		Face Brick Courses
	7 3/4"	
1	8 1/4"	3
2	1' 4 1/2"	6
First brick header	1' 7 1/4"	7
3	2' 3 1/2"	10
4	2' 11 3/4"	13
Second brick header	3' 2 1/2"	14
5	3' 10 3/4"	17
6	4' 7"	20
Third brick header	4' 9 3/4"	21
7	5' 6"	24
8	6' 2 1/4"	27
Fourth brick header	6' 5"	28
9	7' 1 1/4"	31
10	7' 9 1/2"	34
Fifth brick header	8' 0 1/4"	35
15	11' 11"	52
16	12' 7 1/4"	55
Eighth brick header	12' 10"	56

Mortar joint thickness of brick facing and backup: 1/2"

MASONRY—BRICK WALLS (QUANTITY REQUIRED)

Wall Type	Thick	Per 100 Sq. Ft. Wall			Brick Bonding Course
		Block	Brick	Mortar	
4 × 8 × 16 Brick facing	8"	97	772	12 1/2 cu. ft.	7th
8 × 8 × 16 Brick facing	12"	97	868	13 1/2 cu. ft.	7th
8 × 8 × 16 Brick facing	12"	114	788	13 1/2 cu. ft.	6th

MASONRY—BRICKWORK
Percentages to Add to Running Joint Quantities

Common (full header course
    every 5th course)..20%
Common (full header course
    every 6th course) 16 2/3%
Common (full header course
    every 7th course) 14 1/3%
English or English Cross
    (full headers every other course).....................50%
English or English Cross
    (full headers every 6th course)....................16 2/3%
Dutch or Dutch Cross (full
    headers every other course)..........................50%
Dutch or Dutch Cross (full
    headers every 6th course).........................16 2/3%
Flemish (full headers
    every course) 33 1/3%
Flemish (full headers
    every 6th course) 5 2/3%
Double Header (2 headers
    and a stretcher every 6th course) 8 1/3%
Double Header (2 headers
    and a stretcher every 5th course)....................10%

MASONRY—BRICKWORK (WEIGHT)

Wall Thickness (In.)	Wt./ 100 Sq. Ft. Wall
4	3,700
8	7,900
12	11,500
16	15,500

MASONRY—CONCRETE (CALCULATING TABLE)

Number of Block 16 Inches/Course for Solid Walls of Various Sizes

Size in Feet	2	4	6	8	10	12	14	16	18	20	22	24	26	28	30	32	34	36	38	40
2	4	7	10	13	16	19	22	25	28	31	34	37	40	43	46	49	52	55	58	61
4	7	10	13	16	19	22	25	28	31	34	37	40	43	46	49	52	55	58	61	64
6	10	13	16	19	22	25	28	31	34	37	40	43	46	49	52	55	58	61	64	67
8	13	16	19	22	25	28	31	34	37	40	43	46	49	52	55	58	61	64	67	70
10	16	19	22	25	28	31	34	37	40	43	46	49	52	55	58	61	64	67	70	73
12	19	22	25	28	31	34	37	40	43	46	49	52	55	58	61	64	67	70	73	76
14	22	25	28	31	34	37	40	43	46	49	52	55	58	61	64	67	70	73	76	79
16	25	28	31	34	37	40	43	46	49	52	55	58	61	64	67	70	73	76	79	82
18	28	31	34	37	40	43	46	49	52	55	58	61	64	67	70	73	76	79	82	85
20	31	34	37	40	43	46	49	52	55	58	61	64	67	70	73	76	79	82	85	88
22	34	37	40	43	46	49	52	55	58	61	64	67	70	73	76	79	82	85	88	91
24	37	40	43	46	49	52	55	58	61	64	67	70	73	76	79	82	85	88	91	94
26	40	43	46	49	52	55	58	61	64	67	70	73	76	79	82	85	88	91	94	97
28	43	46	49	52	55	58	61	64	67	70	73	76	79	82	85	88	91	94	97	100
30	46	49	52	55	58	61	64	67	70	73	76	79	82	85	88	91	94	97	100	103
32	49	52	55	58	61	64	67	70	73	76	79	82	85	88	91	94	97	100	103	106
34	52	55	58	61	64	67	70	73	76	79	82	85	88	91	94	97	100	103	106	109
36	55	58	61	64	67	70	73	76	79	82	85	88	91	94	97	100	103	106	109	112
38	58	61	64	67	70	73	76	79	82	85	88	91	94	97	100	103	106	109	112	115
40	61	64	67	70	73	76	79	82	85	88	91	94	97	100	103	106	109	112	115	118
42	64	67	70	73	76	79	82	85	88	91	94	97	100	103	106	109	112	115	118	121
44	67	70	73	76	79	82	85	88	91	94	97	100	103	106	109	112	115	118	121	124
46	70	73	76	79	82	85	88	91	94	97	100	103	106	109	112	115	118	121	124	127
48	73	76	79	82	85	88	91	94	97	100	103	106	109	112	115	118	121	124	127	130
50	76	79	82	85	88	91	94	97	100	103	106	109	112	115	118	121	124	127	130	133
52	79	82	85	88	91	94	97	100	103	106	109	112	115	118	121	124	127	130	133	136
54	82	85	88	91	94	97	100	103	106	109	112	115	118	121	124	127	130	133	136	139
56	85	88	91	94	97	100	103	106	109	112	115	118	121	124	127	130	133	136	139	142
58	88	91	94	97	100	103	106	109	112	115	118	121	124	127	130	133	136	139	142	145
60	91	94	97	100	103	106	109	112	115	118	121	124	127	130	133	136	139	142	145	148

MASONRY—CONCRETE (MODULAR SIZES)

Size	Widths (In.)	Lengths (In.)	Heights (In.)	Halves
8 × 6	1 5/8, 2 5/8, 3 5/8 5 5/8, 7 5/8, 11 5/8	15 5/8	7 5/8	Width × 7 5/8 × 7 5/8
8 × 12	3 5/8, 5 5/8, 7 5/8, 9 5/8 11 5/8	11 5/8	7 5/8	Width × 7 5/8 × 5 5/8
5 × 12	3 5/8, 7 5/8	11 5/8	4 7/8	Width × 4 7/8 × 5 5/8

Solid Brick: 3 5/8 × 7 5/8 × 2 1/4 in.
Jumbo Brick: 3 5/8 × 7 5/8 × 3 5/8 in.
Double: 3 5/8 × 7 5/8 × 4 7/8 in.
Half High Block: 3 5/8 × 7 5/8 × 15 5/8 in.
   3 5/8 × 7 5/8 × 11 5/8 in.
Sash Block: Halves: Bullnose made in most types

Size may vary in different localities

MASONRY—CONCRETE (MORTAR BEDDING DIMENSIONS)

Type	Size (In.)	Shell Thickness (In.)	Mortar Bedding Area % of Area	
			Full Area	Face Shell Area
3 – core	8 × 8 × 16	1 3/4	66	51
3 – core	8 × 8 × 16	1 1/2	60	45
3 – core	8 × 8 × 16	1 1/4	52	38
2 – core	8 × 8 × 16	2 1/8	59	54
2 – core	8 × 8 × 16	1 3/4	50	46
2 – core	8 × 5 × 12	1	40	29
2 – core	8 × 3 1/2 × 12	1 3/4	58	50

MASONRY—CONCRETE (MORTAR REQUIREMENTS)
To Lay 1000 Concrete Blocks, 3/8-Inch Joints—Use 1-3 Mix

Size of Block	Thickness of Wall	Cu. Ft. Mortar	Sacks	Cu. Yds. Sand
8 × 8×16	8	40	12.6	1.40
4 × 8×16	4	24.5	7.7	0.85
8 ×12×16	12	50.0	16.0	1.75
8 × 5×12	8	20.0	6.3	0.70

MASONRY—CONCRETE (MORTAR SIZES, WEIGHTS)
For 100 Square Feet of Wall, 3/8-Inch Joints

Unit Sizes	Nominal Wall Thickness (In.)	Number of Units	Average Weight of Finished Wall			For 100 Units
			Heavyweight Aggregate (Lb.)	Lightweight Aggregate (Lb.)	Mortar (Cu. Ft.)	Mortar (Cu. Ft.)
3 5/8 × 3 5/8 × 15 5/8	4	225	4550	3550	13.5	6.0
5 5/8 × 3 5/8 × 15 5/8	6	225	5100	3900	13.5	6.0
7 5/8 × 3 5/8 × 15 5/8	8	225	6000	4450	13.5	6.0
3 5/8 × 7 5/8 × 15 5/8	4	112.5	4050	3000	8.5	7.5
5 5/8 × 7 5/8 × 15 5/8	6	112.5	4600	3350	8.5	7.5
7 5/8 × 7 5/8 × 15 5/8	8	112.5	5550	3950	8.5	7.5
11 5/8 × 7 5/8 × 15 5/8	12	112.5	7550	5200	8.5	7.5

MASONRY—CONCRETE BLOCK

Quick Method of Estimating Concrete Block

To estimate number of concrete block required for a wall, figure the wall into square feet of wall to be laid up, divide that by 8 and multiply by 9, as every 9 blocks lay 8 square feet. For example: you have a house 20×30 ft. with a 7-foot finished basement; 100 feet around the wall 7 feet high equals 700 sq. ft. This figure divided by 8, equals 87.5 times 9 equals 787.5, with a deduction of 5 block for each corner because this wall will be 10 courses high, 767 block less the openings will be needed for this wall.

MASONRY—CONCRETE CHIMNEY BLOCK
(SIZES, WEIGHTS)

Flue Size	Wt. (Lbs.)	Dim. (In.)
8″× 8″	85	15 1/2×15 1/2
7″ – 8″ Rnd.	85	15 1/2×15 1/2
8″×12″	95	16 3/4×21 1/4

MASONRY—CONCRETE PARTITIONS
Load Bearing Partition Limits

Thickness (In.)	Height (Ft.)
2	9
3	12
4	15
6	20
8	25

Check local codes

Masonry—Concrete Walls (Heights), see page 163.

MASONRY—CONCRETE WALLS (LENGTH)

Units No.	Units 15 5/8″ Long and Half Units 7 5/8″ Long with 3/8″ Thick Head Joints	Units 11 5/8″ Long and Half Units 5 5/8″ Long with 3/8″ Thick Head Joints
1	1′ 4″	1′ 0″
1 1/2	2′ 0″	1′ 6″
2	2′ 8″	2′ 0″
2 1/2	3′ 4″	2′ 6″
3	4′ 0″	3′ 0″
3 1/2	4′ 8″	3′ 6″
4	5′ 4″	4′ 0″
4 1/2	6′ 0″	4′ 6″
5	6′ 8″	5′ 0″
5 1/2	7′ 4″	5′ 6″
6	8′ 0″	6′ 0″
6 1/2	8′ 8″	6′ 6″
7	9′ 4″	7′ 0″
7 1/2	10′ 0″	7′ 6″
8	10′ 8″	8′ 0″
8 1/2	11′ 4″	8′ 6″
9	12′ 0″	9′ 0″
9 1/2	12′ 8″	9′ 6″
10	13′ 4″	10′ 0″
10 1/2	14′ 0″	10′ 6″
11	14′ 8″	11′ 0″
11 1/2	15′ 4″	11′ 6″
12	16′ 0″	12′ 0″
12 1/2	16′ 8″	12′ 6″
13	17′ 4″	13′ 0″
13 1/2	18′ 0″	13′ 6″
14	18′ 8″	14′ 0″
14 1/2	19′ 4″	14′ 6″
15	20′ 0″	15′ 0″
20	26′ 8″	20′ 0″

MASONRY—CONCRETE WALLS (HEIGHTS)

3/8-Inch Joints

No. of Courses	Units 7 5/8" High and 3/8" Thick Bed Joint	Units 3 5/8" High and 3/8" Thick Bed Joint
1	8"	4"
2	1' 4"	8"
3	2' 0"	1' 0"
4	2' 8"	1' 4"
5	3' 4"	1' 8"
6	4' 0"	2' 0"
7	4' 8"	2' 4"
8	5' 4"	2' 8"
9	6' 0"	3' 0"
10	6' 8"	3' 4"
15	10' 0"	5' 0"
20	13' 4"	6' 8"
25	16' 8"	8' 4"
30	20' 0"	10' 0"
35	23' 4"	11' 8"
40	26' 8"	13' 4"
45	30' 0"	15' 0"
50	33' 4"	16' 8"

MASONRY—CONCRETE WALLS (THICKNESS)

Wall	RESIDENCES			COMMERCIAL			CAVITY WALLS RESIDENCES			CAVITY WALLS COMMERCIAL		
	One-Story	Two-Story	Three-Story	One-Story	Two-Story	Three-Story	One-Story	Two-Story	Three-Story	One-Story	Two-Story	Three-Story
Basement or Foundation	8"	8"	8"	8"–12"	12"	12"–16"	Solid 8"–10"–12"			Solid 10"–12"		
1st Story	8"	8"	8"	8"–12"	12"	12"–16"	10"	10"	10"–12"	10"–12"	10"–12"	12"
2nd Story		8"	8"		8"–12"	12"		10"	10"		10"–12"	12"
3rd Story			8"			8"–12"			10"			10"–12"

Codes may vary widely

MASONRY—FIREBRICK (QUANTITIES REQUIRED)

Wall Thickness (In.)	Brick/ Sq. Ft.
2 1/2	3 1/2
4 1/2	6 1/4
9	14
13 1/2	21

1 cu. ft. solid brick work - 17 brick
1 cu. ft. solid brick work weighs 136 lbs.

MASONRY—FIREBRICK (SIZES, WEIGHTS)

Type	Size	Wt. (Lbs.)
Std.	4 1/2″ × 9″ × 2 1/2″	9
Split	4 1/2″ × 9″ × 1 1/4″	4 1/2
Arch	4 1/2″ × 9″ × 2 1/2″ × 1 1/8″	7
Wedge	4 1/2″ × 9″ × 2 1/2″ × 1 7/8″	6 1/2

MASONRY—LIGHTWEIGHT CONCRETE (PROPERTIES)
7 5/8 × 15 5/8 Units

Thickness	Average Weight/ Unit Lb.	Average Wt./ Sq. Ft. Wall Lb.	Fire Rating Hours			Decibel Reduction Plaster 2 Sides	U Value	
			Plain Wall	Plaster			Plain	Furred & Plastered
				1 Side	2 Sides			
3″ Solid	16	20	1	1 1/2	2	42	.52	.33
4″ Hollow	17	21	1 1/4	1 3/4	2 1/2	50	.51	.32
4″ Solid	20	26	1 1/2	2	2 1/2	50	.48	.31
6″ Hollow	20	26	1 1/2	2	2 1/2	50	.47	.30
6″ Hollow	25	31	2	2 1/2	3 1/2	51	.35	.24
8″ Hollow	30	37	3	4	5	53	.33	.24
8″ Hollow	34	41	4	5	6	53	.32	.23
10″ Hollow	40	48	5	6	7	55	.31	.23
12″ Hollow	40	48	5	6	7	55	.31	.23

Check local codes and current ratings

MASONRY—SOLID BRICK WALLS
(MORTAR REQUIREMENTS)
100 Square Feet Solid Brick Wall, 1/2-Inch Joints

Cu. Ft. Mortar	4" Wall	8" Wall	12" Wall	16" Wall
	7 1/4	18 1/2	30	41 1/2

1/8" Jts add 21%	5/8" Jts add 5%
1/4" Jts add 14%	3/4" Jts add 10%
3/8" Jts add 7%	7/8" Jts add 15%
	1" Jts add 20%

MASONRY—WALLS & PARTITIONS (WEIGHTS/SQUARE FOOT)
Pounds/Square Foot

Material	Thickness —(In.)					
	2	3	4	6	8	12
Brick	..	..	38	..	80	120
Concrete units - gravel and stone aggregate	..	..	31	44	54	80
- lightweight aggregate	..	..	20	28	34	50
Glass block	..	..	18	..	..	..

MASONRY—WALL TIES

Type	Size	Wt./1000	Packed/Box
Regular	7/8" × 7"	35 lbs.	1000
Cavity or "Z"	6" 5 ga. wire 1/4" × 8"	100 lbs.	250
Corr. Block Dovetail	1/8" × 1" × 5" 1/4" × 5"		
Looped Wire	7" 10 ga. galv. 7" 12 ga. galv.		
Copper	3/4" × 6" 7/8" × 7"	32 oz. 16 oz.	

MASTIC

- Use—wood block floors, etc.
- Package—5 gallons
- Coverage—40 sq. ft./gallon
- Wt./Gallon—8 lbs.

MEASURE—CLOTH

2 1/4 inches	1 nail
4 nails	1 quarter
4 quarters	1 yard

MEASURE—CUBIC

1,728 cubic inches	1 cubic foot
128 cubic feet	1 cord wood
27 cubic feet	1 cubic yard
40 cubic feet	1 ton shpg.
2,150.42 cu. inches	1 standard bushel
268.8 cu. in.	1 standard gallon dry
231 cu. in.	1 standard gallon liquid
1 cubic foot	about 4/5 of a bushel
1 perch	A mass 16 1/2 ft. long, 1 ft. high and 1 1/2 ft. wide, containing 24 3/4 cu. ft.

To get cubic contents—multiply height × width × length.

MEASURE—DRY

2 pints	1 quart
8 quarts	1 peck
4 pecks	1 bushel
36 bushels	1 chaldron

1 Standard U.S. bushel = 2,150.42 cu. ins. 1.2445 cu. ft.
1 British Imperial bushel = 2,218.19 cu. ins. 1.2837 cu. ft.
1 Heaped U.S. bushel = 2,747.715 cu. ins.

Cord of wood: 128 cu. ft.
  Std. size 8' × 4' × 4'

Round baskets - hamper sizes:
  1/8 – 1/4 – 1/2 – 5/8 – 3/4 –
  1 1/4 – 1 1/2 – 2 Bu.

Splint basket sizes:
  4 – 8 – 12 – 16 – 24 – 32 Qts.

MEASURE—LAND (SQUARE TRACTS)

One side of the square tract of land containing:
1/10 acre, is 66 ft. or 4,356 sq. ft.
1/8 acre, is 73.8 ft. or 5,445 sq. ft.
3/8 acre, is 85.2 ft. or 7,260 sq. ft.
1/4 acre, is 104.4 ft. or 10,890 sq. ft.
1/3 acre, is 120.5 ft. or 14,520 sq. ft.
1/2 acre, is 147.6 ft. or 21,780 sq. ft.
3/4 acre, is 180.8 ft. or 32,670 sq. ft.
1 acre, is 208.7 ft. or 43,560 sq. ft.
1 1/2 acres, is 208.7 ft. or 65,340 sq. ft.
2 acres, is 295.2 ft. or 87,120 sq. ft.
2 1/2 acres, is 330. ft. or 108,900 sq. ft.
3 acres, is 361.5 ft. or 130,680 sq. ft.
5 acres, is 466.7 ft. or 217,800 sq. ft.
10 acres, is 660. ft. or 435,600 sq. ft.

A lot 25 feet × 125, contains nearly 1/14 of an acre; 50 feet × 218, 1/4 of an acre.

Dividing the area by one side gives the other side if unknown. Thus a lot, if 25 ft. wide, in order to contain 1/10 of an acre, must be (4356 divided by 25 or 174 1/4 ft. deep).

Measure—Linear, see page 168.

MEASURE—LIQUID

4 gills	1 pint
2 pints	1 quart
4 quarts	1 gallon
1 U.S. gallon	321 cu. ins.
1 British Imperial gallon	277.274 cu. ins., 1.2 U.S. gallon
7.48 U.S. gallons	1 cu. ft.
31 1/2 gallons	1 barrel
2 barrels	1 hogshead

MEASURE—LINEAR

LONG MEASURE

12 inches . . . . .1 foot
3 feet . . . . .1 yard
1760 yards or 5280 feet. . . . . .1 mile
1000 mils . . . . .1 inch
16 ¹/₂ feet or 5 ¹/₂ yards . . . . .1 rod
   (formerly called pole or perch)
40 rods . . . . .1 furlong
8 furlongs . . . . .1 sta. mile
3 miles. . . . . .1 league

French   U.S. and British
1 metre. . . . . .39.37 inches, or 3.28083 feet or 1.09361 yards
0.3048 metre . . . . .1 foot
1 centimetre . . . . .0.3937 inch
2.54 centimetres . . . . .1 inch
25.4 millimetres . . . . .1 inch
1 kilometre . . . . .1093.61 yards or 0.62137 mile

SURVEYOR'S MEASURE

7.92 inches . . . . .1 link
25 links . . . . .1 rod
4 rods. . . . . .1 chain
10 sq. chains or 160 sq. rods . . . . .1 acre
640 acres. . . . . .1 sq. mile
36 sq. miles or 6 miles sq.. . . . . .1 township

METRIC EQUIVALENTS

1 centimeter . . . . .0.3937 in.
1 decimeter. . . . . .3.937 in. or 0.328 ft.
1 meter . . . . .39.37 in. or 1.0936 yds.
1 dekameter . . . . .1.9884 rods
1 kilometer . . . . .0.62137 mile
1 inch . . . . .2.54 centimeters
1 foot . . . . .3.048 decimeters
1 yard . . . . .0.9144 meter
1 rod . . . . .0.5028 dekameter
1 mile . . . . .1.6093 kilometers

MEASURE—SHIPPING

Register ton is used to measure internal capacity of ship. 100 cu. ft. = 1 register ton.

40 cu. ft. = (1 U.S. shipping ton
(32.143 U.S. bushels
(31.16 Imperial bushels

42 cu. ft. = (1 British shipping ton
(33.75 U.S. bushels
(32.719 Imperial bushels

MEASURE—SQUARE

METRIC
1 sq. centimeter — 0.1550 sq. inches
1 sq. decimeter — 0.1076 sq. feet
1 sq. meter — 1.196 sq. yards
1 acre — 3.954 sq. rods
1 hectare — 2.47 acres
1 sq. kilometer — 0.386 sq. miles
1 sq. inch — 6.452 sq. centimeters
1 sq. foot — 9.2903 sq. decimeters
1 sq. yard — 0.8361 square meter
 sq. rod — 0.259 acre
1 acre — 0.4047 hectare
1 sq. mile — 2.59 sq. kilometers

FRENCH
1 square metre — 10.764 sq. ft., or 1.196 sq. yds.
0.836 square metre — 1 sq. yard
6.452 sq. centimeters — 1 sq. inch
1 square millimetre — 0.00155 sq. in. — 1973.5 circular mils
645.2 sq. millimetres — 1 sq. inch
1 centiare = 1 sq. metre — 10.764 sq. ft.
1 are = 1 sq. decametre — 1076.41 sq. ft.
1 hectare = 100 acres — 107641. sq. ft. — 2.4711 acres
1 square kilometre — 0.386109 sq. miles — 247.11 acres

U.S. AND BRITISH

U.S.
144 sq. inches — 1 sq. ft.
9 sq. feet — 1 sq. yd.
30 ¼ sq. yards — 1 sq. rod
40 sq. rods — 1 rod
160 sq. rods — 1 acre
4 rods — 1 acre
640 acres — 1 sq. mile

A square is 100 square feet

MEASURE—WEIGHTS

APOTHECARIES WEIGHT

20 grains	– 1 scruple
3 scruples	– 1 dram
8 drams	– 1 ounce
12 ounces	– 1 pound

AVOIRDUPOIS WEIGHT

27 $^{11}/_{32}$ grains	– 1 dram
16 drams	– 1 ounce
16 ounces	– 1 pound
25 pounds	– 1 quarter
4 quarters	– 1 cwt.
2,000 lbs.	– 1 net or short ton
2,240 lbs.	– 1 gross or long ton

TROY WEIGHT

24 grains	– 1 pwt.
20 pwts.	– 1 ounce
12 ounces	– 1 pound
1 carat	– 3.2 grains (for weighing gold, silver and jewels)

1 gram	– 0.03527 ounce
1 kilogram	– 2.204622 lbs.
1 metric ton	– 0.9842 English ton
1 ounce	– 28.35 grams
1 pound	– 0.4536 kilogram
1 English ton	– 1.0160 metric tons
1 ounce	– 16 drams
1 ounce	– 437.5 grains
1 pound	– 7000 grains

(1 gram is the weight of 1 cubic centimeter of water at 4° Cent.)

French	U.S. and British
1 gramme	– 15.432 grains
0.0648 gramme	– 1 grain
28.35 gramme	– 1 ounce av.
1 kilogramme	– 2.2046 pounds
0.4536 kilogramme	– 1 pound
1 tonne or metric ton	(0.9842 ton of 2240 lbs
or	– (19.68 cwt.
1000 kilogrammes	(2204.6 pounds
1.016 metric tons	(
1016 kilogrammes	– (1 ton of 2240 pounds (

METRIC—UNITS

Prefixes		Meaning		Units
milli –	= one – thousandth	$1/1000$	.001	
centi –	= one – hundredth	$1/100$	.01	"meter" for length
deci –	= one – tenth	$1/10$	.1	
Unit	= one		1	"gram" for weight or mass
deka –	= ten	$10/1$	10	
hecto –	= one hundred	$100/1$	100	"liter" for capacity
kilo –	= one thousand	$1000/1$	1000	

MIRRORS—BEVEL EDGE

- Stock Sizes—24 × 30 – 36 in.
  30 × 30 – 36, 40, 48 in.
- Door Size—16 × 56 in., bevel edge

MIRRORS—TILE

- Sizes (Standard Stock)—12 × 12 in.
  30.5 × 30.5 cm.
- Thickness—1/8 and 1/4 in.
- Pieces/Carton—6
- Availability—Plain and bevel edges

MORTAR—LIME CEMENT
Parts by Volume

Type of Service	Cement	Hydrated Lime or Lime Putty	Mortar Sand in Damp, Loose Condition
For ordinary service	1 – masonry cement or	—	2 to 3
	1 – portland cement	1 to 1 1/4	4 to 6
Subject to extremely heavy loads, violent winds, earthquakes or severe frost action.	1 – masonry cement plus 1 – portland cement or	—	4 to 6
Isolated piers.	1 – portland cement	0 to 1/4	2 to 3

MORTAR—LIME PUTTIES
Quantities of Materials for Lime – Putty Masonry Mortars

| Proportions by Volume | | | Quantity of Materials Required | | | | | |
| | | | For One Cubic Yard of Mortar | | | To Lay 1000 Brick (17 Cu. Ft.) | | |
Cement	Lime Putty	Sand	Cement Bags	Lime Lbs.	Sand Cu. Yd.	Cement Bags	Lime Lbs.	Sand Cu. Yd.
— LUMP QUICKLIME (70 Cu. Ft. Lime Putty per Ton of Lump Quicklime)								
0	1	3	0	257	1	0.	162	.63
1	3	12	2 1/4	193	1	1.42	121 1/2	.63
1	2	9	3	171 1/2	1	1.89	108	.63
1	1 1/2	7 1/2	3.6	154	1	2.27	97	.63
1	1	6	4 1/2	128 1/2	1	2.83	81	.63
1	1/2	4 1/2	6	86	1	3.78	54	.63
1	0	3	9	0	1	5.67	0	.63
1	10%	3	9	26	1	5.67	16	.63
1	15%	3	9	39	1	5.67	24	.63
PULVERIZED QUICKLIME (80 Cu. Ft. Lime Putty per Ton of Pulverized Quicklime)								
0	1	3	0	225	1	0	142	.63
1	3	12	2 1/4	169	1	1.42	106	.63
1	2	9	3	150	1	1.89	94 1/2	.63
1	1 1/2	7 1/2	3.6	135	1	2.27	85	.63
1	1	6	4 1/2	112 1/2	1	2.83	71	.63
1	1/2	6	4 1/2	75	1	3.78	47	.63
1	0	3	9	0	1	5.67	0	.63
1	10%	3	9	22 1/2	1	5.67	14	.63
1	15%	3	9	34	1	5.67	21	.63
HYDRATED LIME (46 Cu. Ft. Lime Putty per Ton of Hydrated Lime)								
0	1	3	0	391	1	0	246	.63
1	3	12	2 1/4	293 1/2	1	1.42	185	.63
1	2	9	3	261	1	1.89	164	.63
1	1 1/2	7 1/2	3.6	235	1	2.27	148	.63
1	1	6	4 1/2	195 1/2	1	2.83	123	.63
1	1/2	4 1/2	6	130 1/2	1	3.78	82	.63
1	0	3	9	0	1	5.67	0	.63
1	10%	3	9	39	1	5.67	25	.63
1	15%	3	9	59	1	5.67	37	.63

MORTAR—PACKAGED MIX

One 80-pound bag will lay:

- 50 bricks (3/8-inch joint)
- 20 blocks (3/8-inch joint)

Or Cover:

- 8 Sq. Ft., 1 inch thick

MORTAR—QUANTITIES
Materials/Cubic Foot

MIXES			QUANTITIES				
Cement (Sack)	Lime (Cu. Ft.)	Sand (Cu. Ft.)	Masonry Cement (Sack)	Portland Cement (Sack)	Lime (Cu. Ft.)	Sand (Cu. Ft.)	Mortar Produced Sack Cement (Cu. Ft.)
1 Masonry cement	–	3	0.33	–	–	0.99	3.0
1 Portland cement	1	6	–	0.16	0.16	0.97	6.2
1 Masonry cement plus 1 Portland cement	–	6	0.16	0.16	–	0.97	6.1
1 Portland cement	1/4	3	–	0.29	0.07	0.86	3.5

MORTAR—QUANTITIES
Materials Needed to Make 1 Cubic Yard

Mix Ratio	Cement Sacks	Lime Lbs.	Sand Cu. Yd.
1:1/10:2	13.00	52	0.96
1:1/10:3	9.00	36	1.00
1:1/10:4	6.75	27	1.00
1:1/4:2	12.70	127	0.94
1:1/4:3	9.00	90	1.00
1:1/4:4	6.75	67	1.00
1:1/2:2	12.40	250	0.92
1:1/2:3	8.80	175	0.98
1:1/2:4	6.75	135	1.00
1:1/2:5	5.40	110	1.00
1:1:3	8.60	345	0.95
1:1:4	6.60	270	0.98
1:1:5	5.40	210	1.00
1:1:6	4.50	180	1.00
1:1 1/2:3	8.10	485	0.90
1:1 1/2:4	6.35	380	0.94
1:1 1/2:5	5.30	320	0.98
1:1 1/2:6	4.50	270	1.00
1:1 1/2:7	3.85	230	1.00
1:1 1/2:8	3.40	205	1.00
1:2:4	6.10	490	0.90
1:2:5	5.10	410	0.94
1:2:6	4.40	350	0.98
1:2:7	3.85	310	1.00
1:2:8	3.40	270	1.00
1:2:9	3.00	240	1.00

MORTAR—QUANTITY YIELD
Prepared Mortar and Sand/Cubic Foot & Cubic Yard Mortar

Mix by Damp Loose Volume	Per Cu. Foot		Per Cu. Yard	
	Cement (Lbs.)	Sand (Lbs.)	Cement (Sacks*)	Sand (Yds.)
1:2	33.2	76	12.8	.95
1:2 1/2	28.3	81	10.9	1.01
1:3	24.6	84	9.5	1.05
1:3 1/2	21.8	87	8.4	1.09

*1 sack weighs 70 lbs.

MORTAR—QUANTITY YIELD

Sacks of Masonry Cement Required to Lay 1000 Bricks

Mix	Thickness of Wall, Inches	Width of Joint, Inches				
		1/4	3/8	1/2	5/8	3/4
1:2	4	2.7	4.1	5.5	7.0	8.4
	8,12,16	4.1	5.6	7.2	8.7	10.2
1:2 1/2	4	2.3	3.5	4.7	6.0	7.2
	8,12,16	3.5	4.8	6.1	7.4	8.7
1:3	4	2.0	3.0	4.1	5.2	6.3
	8,12,16	3.0	4.2	5.3	6.5	7.6
1:3 1/2	4	1.8	2.7	3.6	4.6	5.5
	8,12,16	2.7	3.7	4.7	5.7	6.7

MORTAR BOXES—STEEL

Capacity (Cu. Ft.)	Width (In.)	Length (In.)	Depth (In.)	Weight (Lbs.)
31	42	110	12	190
22	42	86	12	155
14	36	68	12	124
9	28	70	10	79
6	24	53	9	52

MORTAR COLOR—QUANTITIES

Amount of mortar color for 1000 brick laid with 1/2 inch running or common-bond joints

	Regular Strength	
	Lbs.	Qts.
Black, Standard	125 or	80
Black, Double Strength	100 or	70
Buff	100 or	75
Chocolate, Double Strength	100 or	50
Green	100 or	55
Red	125 or	60

N

NAILING SCHEDULE—APA PANEL ROOF SHEATHING

Panel Thickness[b] (In.)	Nailing[c][d]		
	Size	Spacing (In.)	
		Panel Edges	Intermediate
5/16	6d	6	12
3/8	6d	6	12
7/16, 15/32, 1/2	6d	6	12
19/32, 5/8, 23/32, 3/4, 7/8, 1	8d	6	12[a]
1 1/8	8d or 10d	6	12[a]

(a) For spans 48 inches or greater, space nails 6 inches at all supports.

(b) For stapling asphalt shingles to 5/16-inch and thicker panels use staples with a 3/4-inch minimum crown width and a 3/4-inch leg length. Space according to shingle manufacturer's recommendations.

(c) Use common smooth or deformed shank nails with panels to 1 inch thick. For 1 1/8-inch panels, use 8d ring- or screw-shank nails.

(d) Other code-approved fasteners may be used.

NAILING SCHEDULE—FRAMING & SHEATHING (WOOD-FRAME HOUSE)

Joining	Nailing Method	Nails		
		Number	Size	Placement
Header to joist	End-nail	3	16d	
Joist to sill or girder	Toenail	2	10d or 3	8d
Header and stringer joist to sill	Toenail		10d	16 in. on center
Bridging to joist	Toenail each end	2	8d	At each joist
Ledger strip to beam, 2 in. thick		3	16d	
Subfloor, boards:				
1 by 6 in. and smaller		2	8d	To each joist
1 by 8 in.		3	8d	To each joist
Subfloor, plywood:				
At edges			8d	6 in. on center
At intermediate joists			8d	8 in. on center
Subfloor (2 by 6 in., T&G) to joist or girder	Blind-nail (casing) and face-nail		16d	
Soleplate to stud, horizontal assembly	End-nail	2	16d	
Top plate to stud	End-nail	2	16d	
Stud to soleplate	Toenail	4	8d	
Soleplate to joist or blocking	Face-nail		16d	16 in. on center
Doubled studs	Face-nail, stagger		10d	16 in. on center
End stud of intersecting wall to exterior wall stud	Face-nail		16d	16 in. on center
Upper top plate to lower top plate	Face-nail		16d	16 in. on center
Upper top plate, laps and intersections	Face-nail	2	16d	
Continuous header, two pieces, each edge			12d	12 in. on center
Ceiling joist to top wall plates	Toenail	3	8d	
Ceiling joist laps at partition	Face-nail	4	16d	
Rafter to top plate	Toenail	2	8d	

	Nailing method	Number of nails	Nail size	Placement
Rafter to ceiling joist	Face-nail	5	10d	
Rafter to valley or hip rafter	Toenail	3	10d	
Ridge board to rafter	End-nail	3	10d	
Rafter to rafter through ridge board	Toenail	4	8d	
	Edge-nail	1	10d	
Collar beam to rafter:				
2 in. member	Face-nail	2	12d	
1 in. member	Face-nail	3	8d	
1-in. diagonal let-in brace to each stud and plate (4 nails at top)		2	8d	
Built-up corner studs:				
Studs to blocking	Face-nail	2	10d	Each side
Intersecting stud to corner studs	Face-nail		16d	12 in. on center
Built-up girders and beams, three or more members	Face-nail		20d	32 in. on center, each side
Wall sheathing:				
1 by 8 in. or less, horizontal	Face-nail	2	8d	At each stud
1 by 6 in. or greater, diagonal	Face-nail	3	8d	At each stud
Wall sheathing, vertically applied plywood				
3/8 in. and less thick	Face-nail		6d	6 in. edge
1/2 in. and over thick	Face-nail		8d	12 in. intermediate
Wall sheathing, vertically applied fiberboard				
1/2 in. thick	Face-nail			1 1/2 in. roofing nail 3 in. edge and
25/32 in. thick	Face-nail			1 3/4 in. roofing nail 6 in. intermediate
Roof sheathing, boards, 4-, 6-, 8-in. width	Face-nail	2	8d	At each rafter
Roof sheathing, plywood:				
3/8 in. and less thick	Face-nail		6d	6 in. edge and 12 in. intermediate
1/2 in. and over thick	Face-nail		8d	

NAILING SCHEDULE—HARDWOOD STRIP FLOORING
Tongue & Groove Must Be Blind Nailed, Square Edge Face Nailed

Size Flooring	Nails Used	Nail Spacing (Min. 2 Nails/Piece Near Ends)
STRIP TONGUE & GROOVE		
3/4 × 1 1/2, 2 1/4, 3 1/4	2-in. machine driven fasteners 7d or 8d screw or cutnail 2-in 15 gauge staples, 1/2-in. crown	(Blind Nailing) 10-12 in. apart, 8-10 preferred
1/2 × 1 1/2, 2	1 1/2-in. machine driven fasteners 5d screw, cut steel, wire casing nail (install on subfloor)	10 in. apart
3/8 × 1 1/4, 2	1 1/4-in. machine driven fastener 4d bright wire casing nail (install on subfloor)	8 in. apart
SQUARE-EDGE FLOORING		
5/16 × 1 1/2, 2	1-in, 15 gauge fully barbed flooring brad	(Face Nailing) 2 nails every 7 in.
5/16 × 1 1/3	1-in, 15-gauge fully barbed flooring brad	1 nail every 5 in. on alternate sides of strip
PLANK		
3/4 × 3 – 8	2-in machine-driven fasteners, 7d or 8d screw or cut nail 1 3/4-in. length with 3/4-in plywood subfloor on slab	(Blind Nailing) 8 in. apart

Follow manufacturer's instructions for installing plank flooring.
Widths 4 inches and over: Install on subfloor 5/8-inch or thicker plywood or 3/4-inch boards.
On slab use 3/4 inch or thicker plywood.

NAILING SCHEDULE—HARDWOOD STRIP FLOORING

Flooring Dimensions (In.)	Nails	
	Size	Spacing
$25/32 \times 2$, $2\ 1/2$, and $3\ 1/4$ (tongued grooved	8d light flooring nail—use wire or steel-cut casing nail (cut nail is preferable)	10″ apart
$25/32 \times 1\ 1/2$ tongued and grooved)	Same as above	12″ apart
$15/32 \times 1\ 1/2$, and 2 (tongued and grooved)	6d bright wire casing nails	10″ apart
$11/32 \times 1\ 1/2$, and 2 (tongued and grooved)	4d bright wire casing nails	8″ apart
$5/16 \times 1\ 1/2$, and 2 (square-edged)	1 1/8″ barbed-wire flooring brad, No. 16; heads countersunk and puttied	2 nails every 7″

NAILING SCHEDULE—PLYWOOD TO FRAMING

Size Material (In.)	Size Nail
SHEATHING (SUBFLOOR, WALLS, ROOF)	
1/2 or less	6d common or ring-shank
5/8 - 3/4	8d common or 6d ring-shank
7/8 - 1	8d common or ring-shank
1 1/8 - 1 1/4	10d common or 8d ring-shank
SIDING (PANEL)	
1/2 or less	6d stainless steel, hot-dipped galvanized, or aluminum siding or casing
5/8	8d stainless steel, hot-dipped galvanized, or aluminum siding or casing
SUBFLOOR-UNDERLAYMENT	
3/4 or less	6d ring-shank
7/8 - 1	8d ring-shank
1 1/8 - 1 1/4	10d common or 8d ring-shank

Spacing: 6″ on centers at edges, 12″ on centers at intermediate points; floors, 10″ on centers at intermediate points; where spans are 48″, all nails on 6″ centers.

NAILING SCHEDULE—WOOD SHINGLE OR SHAKE

Type	Nail Type	Min. Length (In.)
SHINGLES: NEW ROOF		
16, 18-In. Shingles	Shingle	1 1/4
24-In. Shingles	Shingle	1 1/2
SHAKES: NEW ROOF		
18-In. Straight Split	Box	1 3/4
18, 24-In. Handsplit and Resawn	Box	2
24-In. Tapersplit	Box	1 3/4
18, 24-In. Tapersawn	Box	2

Use 2 Type 304 or 316 Stainless Steel, Hot-Dipped Zinc Coated, or Aluminum Nails/Shingle or Shake.

Nails—Aluminum, see page 182.

NAILS—ANCHOR DOWN

Size	Length (In.)	Gauge	Head (In.)	No./Lb.
2d	1	14	3/16	640
3	1 1/4	14	3/16	528
4	1 1/2	12 1/2	1/4	408
5	1 3/4	12 1/2	1/4	250
6	2	12 1/2	1/4	225
8	2 1/2	11 1/2	9/32	135
10	3	9	9/32	70

NAILS—ALUMINUM

Kind	Length (In.)	Gauge	No./Lb.
Wood	1 7/8	12	604
Siding	2 1/8	11 1/2	468
with	2 3/8	10 1/2	319
Sinker Head	2 7/8	8 1/2	185
Wood	1 7/8	12	604
Siding	2 1/8	11 1/2	468
with	2 3/8	10 1/2	319
Casing Head	2 7/8	8 1/2	185
Cedar	1 1/4	13 1/2	1300
Shake	1 3/4	12 1/2	724
Cedar Shingles	1 1/4	14	1480
Asbestos	1 1/4	13	1230
Siding	1 3/4	12 1/2	720
Asbestos	1 1/4	11 1/2	785
Shingle	1 1/2	11 1/2	659
	1 3/4	11 1/2	544
Insulated	2	11 1/2	495
Siding	2 1/2	10	295
Light Finish			
Insulated	2	11 1/2	495
Siding	2 1/2	10	295
Blk. Alrok Finish			
Gypsum	1 1/8	12 1/2	988
Lath	1 1/4	12 1/2	939
	1 1/2	12	725
Standard	7/8	12 1/2	1313
Shingle	1 1/4	12 1/2	1009
Roofing	3/4	10	746
	7/8	10	663
	1	10	605
	1 1/4	10	491
	1 1/2	10	417
	1 3/4	10	368
	2	10	336
	2 1/2	10	274
Roofing	1 3/4	10	318
with Neoprene	2	10	285
washers attached	2 1/2	10	243
Plain Shank			
Roofing	1 3/4	9 1/4	288
with Neoprene	2	9 1/4	258
washers attached	2 1/2	9 1/4	215
Screw Grip			

NAILS—COMMON WIRE (SIZES, STRENGTH, DIAMETERS)

Size of Nails	Strength in Shear (Lb.)	Diameter (In.)
6d	52	0.113
7d	52	0.113
8d	69	0.131
10d	88	0.148
12d	88	0.148
16d	105	0.162
20d	147	0.192
30d	171	0.207
40d	203	0.225
60d	277	0.263

NAILS—CONCRETE (¼-INCH DIAMETER)

Length (In.)	No./Lb.
1	204
1 3/4	112
2 3/4	60

NAILS—DIMENSIONS

2d	1″
3d	1 1/4″
4d	1 1/2″
5d	1 3/4″
6d	2″
7d	2 1/4″
8d	2 1/2″
9d	2 3/4″
10d	3″
12d	3 1/4″
16d	3 1/2″
20d	4″
30d	4 1/2″
40d	5″
50d	5 1/2″
60d	6″
70d	7″
80d	8″

NAILS—DIMENSIONS

Nail Gauge	Decimal Equivalent	Approx. Dia. (In.)
5 1/2	.200	13/64
7	.177	11/64
9	.1483	5/32
10	.135	9/64
11	.1205	1/8
12	.1055	7/64
12 1/2	.099	3/32
13	.0915	3/32
14	.080	5/64
14 1/2	.076	5/64
16 1/2	.058	1/16

NAILS—EXPANDING DRIVE IN

Shield Size and Length (In.)		Drill Size (In.)	Safe Load (Lbs.)
Dia.	Length		
3/16	× 7/8	3/16	60
3/16	× 1 1/4		100
1/4	× 1	1/4	200
1/4	× 1 1/4		250
1/4	× 1 1/2		300
5/16	× 1 1/4	5/16	250
5/16	× 1 3/4		300
5/16	× 2 1/4		300
5/16	× 2 3/4		300
3/8	× 1 1/2	3/8	400
3/8	× 2		450
3/8	× 3 1/4		500
1/2	× 1	1/2	200
1/2	× 2 1/4		500
1/2	× 2 3/4		550
1/2	× 3 1/2		600

Products and specifications vary

NAILS—MASONRY (SCREW SHANK)

Length (In.)	Gauge	Head Size (In.)	Nails/ Lb.
$1/2$	9	$5/16$	324
$5/8$	9	$5/16$	266
$3/4$	9	$5/16$	230
1	9	$5/16$	176
$1\,1/4$	9	$5/16$	164
$1\,1/2$	9	$5/16$	129
$1\,3/4$	9	$5/16$	108
2	9	$5/16$	95
$2\,1/4$	9	$5/16$	85
$2\,1/2$	9	$5/16$	77
$2\,3/4$	9	$5/16$	70
3	9	$5/16$	62
$3\,1/4$	9	$5/16$	57
$3\,1/2$	8	$3/8$	47
$3\,3/4$	8	$3/8$	44
4	7	$3/8$	35

NAILS—POLE BARN

Size	Length (In.)	Gauge	Head (In.)	Nails/ Lb.
8d	$2\,1/2$	11	$17/64$	117
10d	3	10	$9/32$	78
12d	$3\,1/4$	10	$9/32$	73
16d	$3\,1/2$	9	$5/16$	57
20d	4	7	$3/8$	35
30d	$4\,1/2$	7	$3/8$	30
40d	5	$5\,1/2$	$27/64$	23
40d	5	7	$3/8$	26
50d	$5\,1/2$	$5\,1/2$	$27/64$	21
50d	$5\,1/2$	7	$3/8$	24
60d	6	$5\,1/2$	$27/64$	19
60d	6	7	$3/8$	22
70d	7	$5\,1/2$	$27/64$	16
70d	7	7	$3/8$	20
80d	8	$5\,1/2$	$27/64$	14

NAILS—REQUIREMENTS FOR BUILDING MATERIALS

- 16" No. 1 wood shingles laid 5" to weather require 2 lbs. 3d galv. nails per square.
- Wood shingles on sidewalls require from 1 1/2 to 3 lbs. per square, depending on exposure and grade.
- Asphalt shingles require from 2 to 3 lbs. of 1" roofing nails, galv. depending on style of shingle.
- 48" wood lath requires 6 lbs. 3d fine blued lath nails per 1,000 pieces.
- Metal lath requires 16 lbs. of 2" galv. roofing nails per 100 sq. yds.
- 25/32" exterior insulating sheathing requires 25 lbs. 2" galv. roofing nails per 1,000 sq. ft.
- 3/8" gypsum plaster base requires 9 lbs. 1 1/8" blued nails, 5/16" heads, per 1,000 sq. ft.
- 1/2" insulating plaster base requires 7 lbs. 1 1/4" blued nails, 5/16" heads, per 1,000 sq. ft.

NAILS—REQUIREMENTS FOR PLACING LUMBER

Material	Lumber Size	Nail Size	Lbs./1000 Sq. Ft.
Framing	2 × 4 to 2 × 12	10d com. 20d com.	15 15
Boards, Shiplap, D & M	1 × 4 1 × 6 1 × 8 1 × 10 1 × 12	8d com. 8d com. 8d com. 8d com. 8d com.	45 30 27 22 25
Ceiling	5/8 × 4 1 × 4	6d fin. 8d fin.	8 10
Lap Siding	1/2 × 4 1/2 × 6 1/2 × 8 3/4 × 10	6d box 7d box 7d box 8d box	18 14 10 9

Note: If 1 × 6 D & M flooring is blind, figure half the quantity of nails.

NAILS—ROOFING FASTENERS

Length (In.)	Pieces/Lb.	
	Lead Head	Plain
4	31	59
5	29	47
6	25	39
7	23	34
8	21	30
9	20	27
10	18	24
11	17	21
12	16	20
13	15	17
14	14	15
15	13	13
16	12	11

NAILS—SIZES (COMMON WIRE)

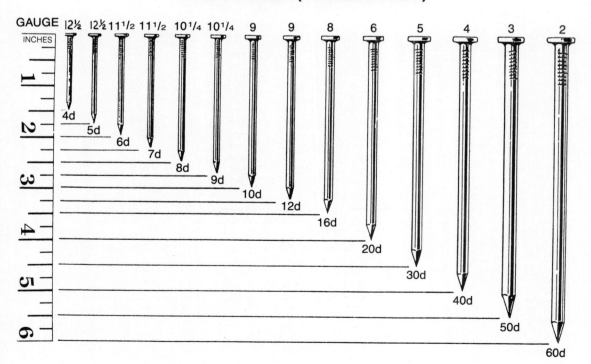

NAILS—SIZES (COMMON WIRE, BRIGHT)

Size	Gauge	Length (In.)	Diameter (In.)
6d	11 1/2	2	0.113
8d	10 1/4	2 1/2	.131
10d	9	3	.148
12d	9	3 1/4	.148
16d	8	3 1/2	.162
20d	6	4	.192
30d	5	4 1/2	.207
40d	4	5	.225
50d	3	5 1/2	.244
60d	2	6	.262

NAILS—SIZES (HELICALLY & ANNULARLY THREADED)

Size	Length (In.)	Diameter (In.)
6d	2	0.120
8d	2 1/2	.120
10d	3	.135
12d	3 1/4	.135
16d	3 1/2	.148
20d	4	.177
30d	4 1/2	.177
40d	5	.177
50d	5 1/2	.177
60d	6	.177
70d	7	.207
80d	8	.207
90d	9	.207

NAILS—SIZES (SMOOTH BOX)

Size	Gauge	Length (In.)	Diameter (In.)
3d	14 1/2	1 1/4	0.076
4d	14	1 1/2	.080
5d	14	1 3/4	.080
6d	12 1/2	2	.098
7d	12 1/2	2 1/4	.098
8d	11 1/2	2 1/2	.113
10d	10 1/2	3	.128
16d	10	3 1/2	.135
20d	9	4	.148

NAILS—SIZES (STANDARD CUT)

Sizes	Lengths (In.)	No./Lb.
3d	1 1/4	432
4d	1 1/2	288
5d	1 3/4	208
6d	2	150
7d	2 1/4	120
8d	2 1/2	88
9d	2 3/4	72
10d	3	65
12d	3 1/4	44
16d	3 1/2	36
20d	4	27
30d	4 1/2	20
40d	5	15
50d	5 1/2	12
CASING FINISH		
6d	2	224
8d	2 1/2	128
10d	3	80

NAILS—WIRE

Type	Size
Barbed Box	2d to 40d
Barbed Roofing	3/4″ to 2″
Blunt Point Shingle	3d to 4d
Blued Lath	2d to 3d
Blued Plaster Board	1″ to 1 3/4″
Common	2d to 60d
Casing	2d to 40d
Common Brads	2d to 60d
Clinch	2d to 20d
Dowel Pin	5/8″ to 2″
Drive Screw	2d to 10d
Drive Screw Floor	5d to 10d
Duplex Head	6d to 30d
Flat Head Spike	Lengths to 16″
Finishing	2d to 20d
Flat Head Barbed Car	4d to 60d
Flooring Brads	6d to 20d
Fetter Ring	3/4″ to 2 1/2″
Flat Head Hinge	4d to 20d
Flooring	6d to 20d
Fiberboard	2d to 5d
Hardened Concrete	1/2″ to 3″
Lath	1 1/8″
Lead Head Roofing	2d to 10d
Oval Head Spike	Various Gauge Lengths to 16″
Oval Head Barbed Car	4d to 60d
Oval Head Hinge	4d to 20d
Siding	5d to 10d
Smooth Box	2d to 40d
Sheet Roofing Fasteners	6″ to 15 1/2″
Smooth Foundry	3/4″ to 3″
Superhead Roofing	3/4″ to 2″
Wood Shingle	1 3/4″

NOSING—LANDING

- Thickness—1 1/6 in.
- Width—3 1/2 in.
- Rabbeted—1/4 × 2 in.
- Lengths—3 ft., 3 ft.-6in., 4ft.

PAINT—AVERAGE COVERING CAPACITY

Material	Square Feet to One Gallon		
	1 Coat	2 Coats	3 Coats
Aluminum Paint	600	425	
Barn Paint	500	275	
Bleaching Solutions	250–300		
Brick Paint – White or Light Tints on Unsurfaced Walls	225	110	75
Brick Paint – Dark Tints on Unsurfaced Walls	290	145	95
Bronzing Liquid	600	425	
Canvas Deck Paint	550	275	
Enamel Base Coat (Prepared)	425	240	165
Enamels	425	215	165
Flat Wall Paint, Dark Colors (on Smooth Finish)	725	365	240
Flat Wall Paint, White or Light Colors (on Rough Sand Finish)	475	265	190
Flat Wall Paint, Dark Colors (on Rough Sand Finish)	625	340	215
Flat Wall Paint Size – Sand Float Finish	300	150	
Flat Wall Paint Size – Hard Finish	500	300	
Flat – White Undercoat	500	300	
Floor Paint – Interior Concrete	350	200	
Floor Paint – Inside	500	275	
Galvanized Iron Primer	700	450	
Glazing Liquid	525		
Graining Colors in Oil (About 1 lb. to 100 Sq. Ft. Reduced with Turpentine)			
Graphite Paint	600	400	
Ground Color	400	225	
Japan (Used in White Paints, Varnishes and Enamels as a drier – about a Tablespoon to a quart. Also used as a size for laying Gold Leaf.)			
Lawn and Porch Furniture Enamel	450	250	
Lead – Flat	600	400	
Linseed Oil	600		
Lacquer	200–300		
Lacquer Sealer	250–300		
Liquid Filler	250–400		
Liquid Wax	600–700		
Mahogany Glaze	600	400	
Non-Grain Raising Stain	275–325		
Oil Stain	300–350		

Material	Square Feet to One Gallon		
	1 Coat	2 Coats	3 Coats
Outside House Paint – White or Light Tints, Porous Woods	475	255	190
Outside House Paint – White or Light Tints, Close Grained Woods	525	275	190
Outside House Paint – Dark Colors, Greys, Tans, etc. Porous Woods	525	280	215
Outside House Paint – Dark Colors, Greys, Tans, etc. Close Grained Woods	575	300	215
Paint and Varnish Remover (1 gallon should remove about 200 Sq. Ft.)			
Paste Wood Filler	40 – 50 (per lb.)		
Pigment Oil Stain	350 – 400		
Paste Wax	125 – 175		
Rubbing Varnish	450 – 500		
Shellac	550	275	
Screen Enamel (For Wood and Wire)	500	300	
Stair, Wood Tints	750		
Stain, Shingle; (2 gals. to 1000 Shingles for dipping 1 coat) (Brushing 1 coat after dipping, 1/2 gal.)			
Structural Steel Paint (Red Lead in Oil)	650	350	
Spirit Stain	250 – 300		
Varnish, Finishing (or Spar) – Soft Wood Trim	475	215	165
Varnish, Finishing (or Spar) – Hard Wood Trim	575	290	215
Varnish, Flat – Soft Wood Floors	425	215	165
Varnish, Flat – Hard Wood Floors	525	265	215
Varnish Stain	550	350	
Waterproof Paint	450	250	
Water Stain	350 – 400		

PAINT—CEMENT (COVERAGE)

- 50 Pounds Dry Comp.
  or
- 5 Gallon Liquid
  will cover:
- 400 Sq. Ft. (1 Coat) Av. Conditions

PAINT—COLOR MIXING

Buff	– White, yellow ochre and red.
Chestnut	– Red, black and yellow.
Chocolate	– Raw umber and red and black.
Claret	– Red, umber and black.
Copper	– Red, yellow and black.
Dove	– White, vermillion, blue and yellow.
Drab	– White, yellow ochre, red and black.
Fawn	– White, yellow and red.
Flesh	– White, yellow ochre and vermillion.
Freestone	– Red, black, yellow ochre and white.
French Grey	– White, Prussian blue and lake.
Grey	– White lead and black.
Gold	– White, stone ochre and red.
Green Bronze	– Chrome green, black and yellow.
Green Pea	– White and chrome green.
Lemon	– White and chrome yellow.
Limestone	– White, yellow ochre, black and red.
Olive	– Yellow, blue, black and white.
Orange	– Yellow and red.
Peach	– White and vermillion.
Pearl	– White, black and blue.
Pink	– White, vermillion and lake.
Purple	– Violet, with more red and white.
Rose	– White and madder lake.
Violet	– Red, blue and white.

Paint—Drying Time, see page 194.

PAINT—REFLECTION VALUES

Paint Colors	%	Paint Colors	%
Apple Green	31%	Light Buff	73%
Black	2%	Light Cream	77%
Bright Sage	52%	Light Orchid	79%
Buff	63%	Light Gray	52%
Caen Stone	76%	Orchid	67%
Canary Yellow	77%	Olive Tan	43%
Cocoanut Brown	16%	Oyster White	77%
Coral	64%	Pale Green	59%
Cream	77%	Pastel Green	73%
Cream Gray	66%	Sea Green	67%
Forest Green	22%	Shell Pink	55%
French Blue	47%	Silver Gray	46%
Ivory	82%	Sky Blue	65%
Ivory Tan	66%	Sunlight Yellow	77%
Light Brown	20%	White	89%

PAINT—DRYING TIME

Material	Touch	Recoat	Rub
Lacquer	1 – 10 min.	1 1/2 – 3 hrs.	16 – 24 hrs.
Lacquer Sealer	1 – 10 min.	30 – 45 min.	1 hr. (sand)
Paste Wood Filler	. . . .	24 – 48 hrs.	. . . .
Paste Wood Filler (Q.D.)	. . . .	3 – 4 hrs.	. . . .
Water Stain	1 hr.	12 hrs.	. . . .
Oil Stain	1 hr.	24 hrs.	. . . .
Spirit Stain	Zero	10 min.	. . . .
Shading Stain	Zero	Zero	. . . .
Non-Grain Raising Stain	15 min.	3 hrs.	. . . .
NGR Stain (Quick-Dry)	2 min.	15 min.	. . . .
Pigment Oil Stain	1 hr.	12 hrs.	. . . .
Pigment Oil Stain (Q.D.)	1 hr.	3 hrs.	. . . .
Shellac	15 min.	2 hrs.	12 – 18 hrs.
Shellac (Wash Coat)	2 min.	30 min.	. . . .
Varnish	1 1/2 hrs.	18 – 24 hrs.	24 – 48 hrs.
Varnish (Q.D.) Synthetic	1/2 hr.	4 hrs.	12 – 24 hrs.

PAINT—REQUIREMENTS FOR INTERIORS & EXTERIORS

On interior work, for rough, sand-finished walls or unpainted wallboard, add 50% to quantities; for each door or window deduct 1/2 pint of materials for walls. For trim, add 1/8 to 1/5 of the amount required for the body. For exterior blinds, 1/2 gallon will cover 12 to 14 blinds, one coat.

Room Perimeter	Ceiling 8 Feet	Ceiling 8 1/2 Ft.	Ceiling 9 Feet	Ceiling 9 1/2 Ft.	Paint for Ceiling	Finish for Floors	Each Door or Window
30 feet	5/8 gal	5/8 gal.	3/4 gal.	3/4 gal.	1 Pt.	1 Pt.	Each window and frame requires 1/4 pint
35	3/4	3/4	3/4	7/8	1 Qt.	1 Pt.	
40	7/8	7/8	7/8	1	1 Qt.	1 Qt.	
45	7/8	1	1	1 1/8	3 Pts.	1 Qt.	
50	1	1 1/8	1 1/8	1 1/4	3 Pts.	1 Qt.	Each door and frame requires 1/2 pint
55	1 1/8	1 1/8	1 1/4	1 1/4	2 Qts.	3 Pts.	
60	1 1/4	1 1/4	1 3/8	1 3/8	2 Qts.	3 Pts.	
70	1 3/8	1 1/2	1 1/2	1 5/8	3 Qts.	2 Qts.	
80	1 1/2	1 5/8	1 3/4	1 7/8	1 Gal.	5 Pts.	

House Perimeter	Av. Ht. 12 Feet	Av. Ht. 15 Feet	Av. Ht. 18 Feet	Av. Ht. 21 Feet	Av. Ht. 24 Feet
60 feet	1 gal.	1 1/4 gal.	1 1/2 gal.	1 3/4 gal.	2 gal.
76	1 1/4	1 1/2	2	2 1/4	2 1/2
92	1 1/2	2	2 1/2	2 3/4	3
108	1 3/4	2 1/4	2 3/4	3 1/4	3 3/4
124	2	2 1/2	3 1/4	3 3/4	4 1/4
140	2 1/2	3	3 1/2	4	4 1/2
156	2 3/4	3 1/4	4	4 1/2	5 1/4
172	3	3 3/4	4 1/2	5	5 3/4

PARTICLEBOARD—APPLICATIONS BY GRADE

INTERIOR

Application	Grade	Product References
Floor Underlayment	1-M-1	ICBO,SBCC,BOCA, One- and Two-Family Dwelling Code
Mobile Home Decking	Class D-2 Class D-3	NPA 1-82 HUD-Mobile Home Construction and Safety Standards
Shelving	1-M-1 1-M-2 1-M-3	
Counter Tops	1-M-2 1-M-3	ANSI A 161.2
Kitchen Cabinets	1-M-1 1-M-2	ANSI A 161.1
Door Core	1-LD-1 1-LD-2 1-M-3	N.W.W.D.A. Industry Standard Series I.S. 1-78 (Wood Flush Doors) HUD/FHA UM 70a ICBO ER 3390
Stair Treads	1-M-3	
Mouldings	1-M-3	W.W.M.M.P. Standard WM 2-73

Grades shown refer to ANSI A208.1 except for Mobile Home Decking which refers to NPA 1-82.
See Research Reports of Code Agencies for individual companies.

EXTERIOR

Application	Grade	Product References
Roof Sheathing	2-M-W	ICBO,SBCC,BOCA, One- and Two-Family Dwelling Code
Wall Sheathing	2-M-1	ICBO,SBCC,BOCA, One- and Two-Family Dwelling Code
Wall Sheathing	2-M-W	ICBO,SBCC,BOCA, One- and Two-Family Dwelling Code
Combined Sub-Floor Underlayment	2-M-3 2-M-W	ICBO,SBCC,BOCA, One- and Two-Family Dwelling Code
Siding	2-M-1	ICBO,SBCC,BOCA, One- and Two-Family Dwelling Code
Siding	2-M-W	SBCC,BOCA

Grades shown refer to ANSI A208.1
See Research Reports of Code Agencies for individual companies.

PARTICLEBOARD—MAT-FORMED GRADES & PROPERTIES (TYPE 1)[a]

(Average Values for Sample Consisting of 5 Panels)[b]

| Grade | Length and Width Tolerance (Inches) | Thickness Tolerance | | Modulus of Rupture (psi) | Modulus of Elasticity (psi) | Internal Bond (psi) | Hardness (Pounds) | Linear Expansion Max. Avg. (Percent) | Screwholding | |
		Panel Average[d] (Inches)	Within Panel[e] (Inches)						Face (Pounds)	Edge (Pounds)
1-H-1	± 1/16	± 0.010	± 0.005	2400	350,000	130	500	NS[c]	400	300
1-H-2	± 1/16	± 0.010	± 0.005	3000	350,000	130	1000	NS	425	350
1-H-3	± 1/16	± 0.010	± 0.005	3400	400,000	140	1500	NS	450	350
1-M-1	+ 0 − 1/8	± 0.015	± 0.010	1600	250,000	60	500	0.35	NS	NS
1-M-2	± 1/16	± 0.010	± 0.005	2100	325,000	60	500	0.35	225	200
1-M-3	± 1/16	± 0.010	± 0.005	2400	400,000	80	500	0.35	250	225
1-LD-1	± 1/16	+ 0.005 −0.015	± 0.005	400	80,000	20	NS	0.35	90	NS
1-LD-2	+ 1/16	+ 0.005 −0.015	± 0.005	800	150,000	20	NS	0.35	125	NS

(a) Made with urea-formaldehyde resin binders or equivalent bonding systems.
(b) Except for dimensional tolerances which are individual panel values.
(c) NS—Not Specified
(d) Panel Average from Nominal
(e) Individual Measurement from Panel Average

PARTICLEBOARD—MAT-FORMED GRADES & PROPERTIES (TYPE 2)[a]

(Average Values for Sample Consisting of 5 Panels)[b]

Grade	Length and Width Tolerance (Inches)	Thickness[e] Tolerance		Modulus of Rupture (psi)	Modulus of Elasticity (psi)	Internal Bond (psi)	Hardness (Pounds)	Linear Expansion Max. Avg. (Percent)	Screwholding	
		Panel Average[g] (Inches)	Within Panel[h] (Inches)						Face (Pounds)	Edge (Pounds)
2-H-1	± 1/16	± 0.015	± 0.005	2400	350,000	125	500	NS[d]	400	300
2-H-2	± 1/16	± 0.015	± 0.005	3400	400,000	300	1800	NS	450	350
2-M-1	+0 – 1/8	± 0.015	± 0.010	1800	250,000	60	500	0.35	225	160
2-M-2	+0 – 1/8	± 0.015	± 0.010	2500	450,000	60	500	0.35	250	200
2-M-3	+0 – 1/8	± 0.015	± 0.010	3000	500,000	60	500	0.35	NS	NS
2-M-W[c]	+0 – 1/8	± 0.015	± 0.010	2500	450,000	50	500	0.20	NS	NS
2-M-F[f]	+0 – 1/8	± 0.015	± 0.010	3000	500,000	50	500	0.20	NS	NS

(a) Made with Phenol-formaldehyde resins or equivalent bonding systems.
(b) Except for dimensional tolerances which are individual panel values.
(c) "W" indicates that this product is made from wafers
(d) NS—Not Specified
(e) Thickness tolerance values shown are for sanded panels as defined by the manufacturer. Values for unsanded panels for all 2-M Grades shall be ± 0.030 for panel average and ± 0.030 for within panel
(f) "F" indicates that this product is made from flakes
(g) Panel average from nominal
(h) Individual measurement from panel average

PARTICLEBOARD SHELVING—DIMENSIONS FOR UNIFORM LOADING

END SUPPORTED MAXIMUM SPAN

Load* Inches	Underlayment 1-M-1			Industrial Particleboard 1-M-2			MDF		
	$1/2$	$5/8$	$3/4$	$1/2$	$5/8$	$3/4$	$1/2$	$5/8$	$3/4$
50.0	14	18	21	15	19	23	15	19	23
45.0	15	18	22	16	20	24	16	19	23
40.0	15	19	23	17	21	25	16	20	24
35.0	16	20	24	17	22	26	17	21	25
30.0	17	21	25	18	23	27	18	22	27
25.0	18	22	26	19	24	29	19	23	28
20.0	19	24	28	21	26	31	20	25	30
17.5	20	25	29	22	27	32	21	26	31
15.0	21	26	31	23	28	34	22	27	33
12.5	22	27	32	24	30	35	23	29	34
10.0	23	29	34	26	32	38	25	31	37
7.5	25	31	37	28	34	40	27	33	39
5.0	28	34	41	31	38	44	30	37	43

OVERHANGING SHELF MAXIMUM OVERHANG

Load* Inches	Underlayment 1-M-1			Industrial Particleboard 1-M-2			MDF		
	$1/2$	$5/8$	$3/4$	$1/2$	$5/8$	$3/4$	$1/2$	$5/8$	$3/4$
50.0	6	8	10	7	9	11	7	9	10
45.0	7	8	10	7	9	11	7	9	11
40.0	7	9	11	8	10	12	7	9	11
35.0	7	9	11	8	10	12	8	10	12
30.0	8	10	12	8	10	13	8	10	12
25.0	8	10	12	9	11	13	9	11	13
20.0	9	11	13	10	12	14	9	12	14
17.5	9	11	14	10	12	15	10	12	14
15.0	10	12	14	10	13	16	10	13	15
12.5	10	13	15	11	14	16	11	13	16
10.0	11	13	16	12	15	17	11	14	17
7.5	12	14	17	13	16	19	12	15	18
5.0	13	16	19	14	17	21	14	17	20

MULTIPLE SUPPORTS MAXIMUM SPAN

Load* Inches	Underlayment 1-M-1			Industrial Particleboard 1-M-2			MDF		
	$1/2$	$5/8$	$3/4$	$1/2$	$5/8$	$3/4$	$1/2$	$5/8$	$3/4$
50.0	19	24	29	21	26	31	20	25	30
45.0	20	25	30	22	27	32	21	26	32
40.0	21	26	31	22	28	34	22	27	33
35.0	21	27	32	23	29	35	23	28	34
30.0	23	28	34	25	31	37	24	30	36
25.0	24	30	36	26	33	39	25	32	38
20.0	26	32	38	28	35	42	27	34	40
17.5	27	33	40	29	36	43	28	35	42
15.0	28	35	41	31	38	45	30	37	44
12.5	29	37	44	32	40	48	31	39	46
10.0	31	39	46	34	43	50	33	41	49
7.5	34	42	50	37	46	54	36	45	53
5.0	38	46	55	41	51	60	40	49	58

PERLITE—EXPANDED

Plaster and Concrete Aggregate:

- Weight—7 1/2 – 15 lbs./Cu. Ft.
- Packed—3 and 4 Cu. Ft. Bags

PERLITE—EXPANDED (K FACTOR)

Bulk Density Expanded Perlite

Pounds/Cubic Foot	K Factor
5	0.27
6	0.325
8.5	0.345
11	0.450

PERMEANCE—COMMON VAPOR BARRIER MATERIALS

Material	Permeance
Aluminum foil	0.0
Polyethylene sheet 6-mil	0.08
Aluminum paint	0.5
Latex vapor barrier paint, (1 coat)	0.6
Vinyl wallpaper, heavy	1.0
Oil paint on plaster, (2 coats)	2.0
Latex paint on wood, (3 coats)	10.0
Uncoated wallpaper	20.0

PERMEANCE— POLYETHYLENE FILM

Thickness (In.)	Perms
.001	.67
.002	.33
.003	.22
.004	.17
.005	.13
.006	.11
.007	.095
.008	.08
.009	.075
.010	.07

PERMEANCE—SOFTWOOD PLYWOOD

Values tabulated below (in perms) represent the average water vapor transmitted through plywood in grains per square foot, per hour, per inch of mercury pressure.

Product	Surface Finish	Water Vapor Permeance in Grains/Sq. Ft./Hour/In. Hg (Perms)
Interior-type Plywood 1/4″	None	1.9
Interior-type Plywood 1/4″	Two coats of inside flat wall paint	1.1
Interior-type Plywood 1/4″	One natural blonde thin under-coat plus one coat shellac	0.8
Interior-type Plywood Sheathing 5/16″	None	1.8
Exterior-type Plywood 3/8″[1]	None	0.8
Exterior-type Plywood 3/8″	One coat exterior oil primer plus two coats exterior finish paint	0.2
Exterior Medium Density Overlaid Plywood (1 Side) 3/8″	None	0.3
Exterior Medium Density Overlaid Plywood (2 Sides) 1/2″	None	0.2
Exterior High Density Overlaid Plywood (2 Sides) 1/2″ & 5/8″	None	0.1
FRP[2]/Plywood 3/4″	None	< 0.1

1 Range of seven species: 0.5 – 1.4
2 Fiberglass – Reinforced – Plastic – overlaid plywood

Permeance—Typical, see page 202.

PIPE—ALUMINUM STANDARD

Size (In.) L.P.S.	Outside Diameter (In.)	Inside Diameter (In.)	Wall Thickness (In.)	Appr. Wt. (Lbs.)/Foot
1/2	.840	.622	.109	.294
3/4	1.050	.824	.113	.390
1	1.315	1.049	.133	.580
1 1/4	1.660	1.380	.140	.785
1 1/2	1.900	1.610	.145	.939
2	2.375	2.067	.154	1.262

PERMEANCE—TYPICAL

Materials	Loss in Grs./Hour
Foil surface insulation (2 faces)	.061 – 0.09
Roll roofing – smooth surface	.093 – .12
Asphalt impregnated and surface coated sheathing paper glossy surfaced	
50# 500 sq. ft. roll	.15 – .555
35# 500 sq. ft. roll	.123 – 1.48
Duplex of laminated papers 30 – 30 – 30	.990 – 1.85
Duplex of laminated papers 30 – 60 – 30	.370 – .617
Duplex papers reinforced	.493 – 1.48
Duplex paper coated with metal oxides	.37 – .93
Insulation backup paper, treated	.617 – 2.462
Gypsum lath with aluminum foil backing	.061 – .277
Plaster – wood lath	7.90
Plaster – 3 coats lead and oil	2.650 – 2.77
Plaster – 3 coats flat wall paint	3.08
Plaster – 2 coats aluminum paint	.831
Plaster – fiberboard or gypsum lath	14.20 – 14.8
Slaters felt	3.70 – 18.5
Plywood – 1/4″ Douglas fir, soy bean glue, plain	3.08 – 4.62
2 coats asphalt paint	.308
2 coats aluminum paint	.93
1/2″ 5 – ply Douglas fir	1.920 – 1.975
1/4″ 3 – ply Douglas fir, art. resin glue	3.080 – 4.62
1/2″ 5 – ply Douglas fir, art. resin glue	1.975 – 2.42
Insulating lath & sheathing – board type	18.50 – 24.65
Insulating sheathing, surface coated	2.19 – 3.05
3/16″ compressed fiberboard	3.64
1″ insulating cork blocks	4.44
1/2″ and 1″ blanket insulation between coated papers	1.380 – 1.44
4″ mineral wool – unprotected	20.95

NOTE: These results are given in terms of loss in grains per sq. ft. per hour. The higher the loss, the greater the vapor permeability.

PIPE—BLACK

Size (In.)	Weight (Lbs./Ft.)
1/4	.429
3/8	.570
1/2	.812
3/4	1.067
1	1.616
1 1/4	2.133
1 1/2	2.746
2	3.500
2 1/2	5.740
3	7.560
3 1/2	9.020
4	10.580
5	15.350
6	19.800

PIPE—CONCRETE
Bell End Reinforced—4-Inch Lengths

Inside Dia. (In.)	Wall Thickness (In.)	Wt./ Lin. Ft. (Lbs.)
STANDARD TYPE		
12	2	100
15	2	125
18	2 1/4	160
21	2 3/8	205
24	2 1/2	245
27	2 3/4	300
30	3	370
36	3 1/2	510
42	3 3/4	660
48	4 1/4	835
EXTRA STRENGTH		
15	2 1/4	150
18	2 1/2	205
21	2 3/4	255
24	3	320
30	3 1/2	470
36	4	600
42	4 1/2	750
48	5	1000

PIPE—CONCRETE (SIZES)

Pipe	Sizes (In.)	Joints
Pl. Sewer Pipe	6 to 24	3' long
Pl. Sewer Pipe Xtra Str.	6 to 18	3' long
Pl. Sewer Pipe Reinf.	12 to 96	
Culvert Pipe Reinf.	6 to 120	
Culvert Pipe Reinf. Xtra	24 to 120	

PIPE—CONCRETE & DRAIN TILE
Plain Type Reinforced—Sizes, Weights

Inside Dia. (In.)	Length	Wall Thickness	Wt.(Lbs.)/ Lin. Ft.
PIPE			
6	3	1	25
8	3	1	35
10	3	1 1/8	48
12	4	1 1/2	70
15	4	1 3/4	110
18	4	2	140
21	4	2	205
24	4	2 1/2	245
DRAIN TILE			
6	3	1	20
8	3	1	30
10	3	1 1/8	40
12	3	1 1/4	50
15	3	1 1/4	60
18	3	1 1/2	93
24	4	2	160

PIPE—CONCRETE
(TONGUE & GROOVE REINFORCED)

Inside Dia. (In.)	Laying Length (Ft.)	Wall Thickness (In.)	Wt.(Lbs.)/Lin. Ft.
6	3	1 3/4	48
8	4	2	65
10	4	2	80
12	4	2	88
15	4	2 1/4	125
18	4	2 1/2	160
21	4	2 3/4	205
24	4	3	260
27	4	3 1/4	310
30	4	3 1/2	370
33	4	3 3/4	450
36	4	4	520
39	4	4 1/4	600
42	4	4 1/2	680
45	4	4 3/4	760
48	4	5	850
54	4	5 1/2	1050
60	4	6	1280
66	4, 5, 6	6 1/2	1480
72	4, 5, 6	7	1835
78	4, 5, 6	7 1/2	2150
84	4, 5, 6	8	2300
90	4, 5, 6	8	2600
96	4, 5, 6	8 1/2	2750
102	4, 5, 6	8 1/2	3050
108	4, 5, 6	9	3450

PIPE—CONCRETE
(ELLIPTICAL, SIZES & WEIGHTS)
Lengths: 18 Inches – 4 Feet, All Others 6 Feet

Equiv. Size (In.)	Wall Thickness (In.)	Lbs./Lineal Ft.
18	2 3/4	195
24	3 1/4	300
27	3 1/2	365
30	3 3/4	430
33	3 3/4	475
36	4 1/2	625
39	4 3/4	720
42	5	815
48	5 1/2	1000
54	6	1235
60	6 1/2	1475
66	7	1745
72	7 1/2	2040
78	8	2350
84	8 1/2	2680
90	9	3050
96	9 1/2	3420
102	9 3/4	3725
108	10	4050
114	10 1/2	4470
120	11	4930
132	12	5900
144	13	7000

PIPE—CONDUCTOR
(ROOF LEADERS, TYPES & SIZES)

Type	Size (In.)	Area (Sq. In.)
Smooth Round	3	7.07
	4	12.57
	5	19.63
	6	28.27
Corr. Round	3	5.94
	4	11.04
	5	17.72
	6	25.97
Plain Sqr.	1 3/4 × 2 1/4	3.94
	2 × 3	6
	2 × 4	8
	3 × 4	12
	4 × 5	20
	4 × 6	24
Corr. Sqr.	2 – (1 3/4 × 2 1/4)	3.80
	3 – (2 3/8 × 3 1/4)	7.73
	4 – (2 3/4 × 4 1/4)	11.70
	5 – (3 3/4 × 5)	18.75

All packed 250 ft. (25 pcs) per Bdl.
All 26 – 28 ga. 10′ lengths

PIPE—
COPPER
(STANDARD)

Pipe Size (In.)	Wt.(Lbs.)/ Lin. Ft.
1/8	.259
1/4	.457
3/8	.641
1/2	.955
3/4	1.30
1	1.82
1 1/4	2.69
1 1/2	3.20
2	4.22
2 1/2	6.12
3	8.75
3 1/2	11.4
4	12.9
5	16.2
6	19.4
8	31.6
10	46.2
12	56.5

PIPE—COPPER (WATER CAPACITY)

Delivery Gallons/Minute (Up to 15 Foot Lengths). Multiply ×3 to Get Delivery through 100 Foot Lengths

3/8"	1/2"	5/8"	3/4"	1"	1 1/4"	1 1/2"	2"	2 1/2"	3"	Pressure Drop (Lbs.)
3	6	10	14	35	70	100	200	400	700	5
4	10	17	25	50	100	150	325	600	1000	10
6	14	24	35	75	150	215	500	900	1500	20
8	16	30	45	90	180	275	600	1100	1800	30
9	18	36	55	100	220	310	700	1300	2200	40
10	20	42	65	110	240	350	800	1500	2500	50

PIPE—CULVERT (CORRUGATED METAL)

Recording Gauges – Various Fill Conditions

Dia. (In.)	HIGHWAY WORK				RAILWAY WORK			
	Fills Up to 10'	Fills 10–20'	Fills 20–30'	Fills 30–40'	Fills Up to 10'	Fills 10–20'	Fills 20–30'	Fills 30–40'
8	16 ga.	16 ga.	16 ga.	16 ga.	—	—	—	—
10	16	16	16	16	—	—	—	—
12	16	16	16	16	—	—	—	—
15	16	16	16	16	—	—	—	—
18	16	16	14	14	14 ga.	14 ga.	14 ga.	14 ga.
21	16	16	14	14	14	14	14	14
24	14	14	14	14	14	14	14	14
30	14	14	12	12	14	14	12	12
36	14	12	12	10	12	12	12	10
42	12	12	10	8	12	12	10	8
48	12	10	8	8	10	10	8	8
54	12	8	8	8	8	8	8	8
60	10	8	8	8	8	8	8	8
66	10	8	8	8	—	—	—	—
72	10	8	8	8	—	—	—	—
78	8	8	8	8	—	—	—	—
84	8	8	8	8	—	—	—	—

PIPE—CULVERT (CORRUGATED METAL— SIZES, AREAS WEIGHTS)

Diameter (In.)	Area (Sq. Ft.)	Weight In Pounds/Foot				
		Gauge Metal				
		16	14	12	10	8
8	0.349	7.3	–	–	–	–
10	0.544	9.0	11.0	–	–	–
12	0.785	10.5	12.9	17.9	–	–
15	1.227	12.9	15.7	21.8	–	–
18	1.767	15.3	18.8	26.1	–	–
21	2.405	17.7	22.0	30.4	–	–
24	3.142	20.0	25.2	34.3	43.8	53.2
30	4.909	–	30.9	42.7	54.4	66.2
36	7.068	–	36.7	51.0	64.9	78.9
42	9.621	–	42.6	59.5	75.4	91.7
48	12.566	–	48.9	68.0	86.4	105.1
54	16.000	–	55.7	77.8	98.4	119.6
60	19.635	–	–	85.5	108.9	132.4
66	23.758	–	–	93.7	119.4	145.1
72	28.274	–	–	101.9	130.4	157.9
78	33.183	–	–	–	140.3	170.6
84	38.485	–	–	–	150.8	185.2

PIPE—CULVERT (CORRUGATED METAL)

Unsupported Spans, Elevated Pipe Full of Water

Dia. (In.)	Gauge	Maximum Span (Ft.)
8	16	11
10	16	11
12	16	12
15	16	13
18	16	13
21	16	14
24	14	15
30	14	16
36	12	17
42	12	18
48	12	18
54	12	19
60	10	20
66	10	20
72	10	21
78	8	22
84	8	22

PIPE—CULVERT
(METAL, SIZES & WEIGHTS)

Sizes (In.)	Ga.	Weights (Lbs.)
8	16	7
10	16	9
12	16	11
15	14 – 16	14
18	14 – 16	16
21	14 – 16	18
24	12 – 14 – 16	22
30	12 – 14	32 (14 ga.)
36	8 – 10 – 12 – 14	37
42	10 – 12	44
48	10 – 12	51
54	10 – 12	
60	10 – 12	
66	10 – 12	
72	10 – 12	
84	8 – 10	

PIPE—FURNACE (AREAS, GRILLES, DUCTS)

Pipe Size (In.)	Free Area (In.)	Grille Size (In.)	Duct Sizes, if Used (In.)		
14	154	8 × 30	(20 × 8)	(16 × 10)	(14 × 12)
16	201	10 × 30	(26 × 8)	(20 × 10)	(18 × 12)
18	254	12 × 30	(32 × 8)	(26 × 10)	(22 × 12)
20	314	14 × 30	(40 × 8)	(32 × 10)	(26 × 12)
22	380	18 × 30	(48 × 8)	(38 × 10)	(32 × 12)
24	452	20 × 30	(46 × 10)	(38 × 12)	
26	531	24 × 30	(54 × 10)	(44 × 12)	
28	616	Two 14 × 30	(62 × 10)	(52 × 12)	
30	707	Two 16 × 30	(60 × 12)		

PIPE—FURNACE
(DUCT SIZES)

Sizes (In.)	Lengths (In.)
19 × 8 to 36 × 8	16 – 32
4 × 8 to 18 × 8	16 – 32

Other sizes available

PIPE—FURNACE (ROUND)

Round Sizes (In.)	Ga.	Lengths (In.)	Pckd.
3 – 4 – 5 – 6 – 7 – 8 – 9 – 10 – 12 – 14 – 16 – 18 – 20 – 22 – 24.	24	24	10 Jts.
3 – 4 – 5 – 6 – 7 – 8 – 9 – 10 – 12 – 14 – 16 – 18 – 20 – 22 – 24.	26	24	10 Jts.
3 – 4 – 5 – 6 – 7 – 8 – 9 – 10 – 12	30	24	10 Jts.

Other sizes available

PIPE—LINES (WATER CAPACITY)

Cubic Feet/Second, 1 Foot/1000 = Foot Slope

Inside Dia. (In.)	Water Area	Corr. Metal Pipe	Clay Pipe or Brick Lined	Concrete Pipe
6	0.02 sf	0.11	0.15	0.18
8	0.03	0.24	0.34	0.39
10	0.05	0.04	0.6	0.7
12	0.08	0.07	1.0	1.1
15	1.2	1.3	1.8	2.1
18	1.8	2.1	2.8	3.4
21	2.4	3.1	4.3	5.0
24	3.1	4.4	6.1	7.2
27	4.0	6.1	8.6	9.9
30	4.9	8.1	11.3	13.1
36	7.1	13	18	21
42	9.6	20	28	32
48	12.6	28	40	46
54	15.9	38	52	62
60	19.6	51	70	83
66	23.8	66	92	107
72	28.3	83	118	134
84	38.5	126	175	204
96	50.3	179	250	290
108	63.6	247	340	400

PIPE—SEWER (COMPARATIVE SIZES & WEIGHTS)

Diameter of Pipe (In.)	CORRUGATED METAL PIPE			DOUBLE STRENGTH VITRIFIED CLAY PIPE		PLAIN CONCRETE SEWER PIPE		REINFORCED CONCRETE PIPE – BELL END	
	Gauge	Wall Thickness (In.)	Wt./Lin. Ft. (Lbs.)	Wall Thickness (In.)	Wt./Lin. Ft. (Lbs.)	Wall Thickness (In.)	Wt./Lin. Ft. (Lbs.)	Wall Thickness (In.)	Wt./Lin. Ft. (Lbs.)
8	16	0.06	7.3	0.81	24	1.125	40		
10	16	0.06	9.0	0.88	33	1.25	50		
12	16	0.06	10.5	1.00	45	2.00	106	2.00	100
15	16	0.06	12.9	1.25	70	2.00	128	2.25	150
18	16	0.06	15.3	1.50	100	2.00	155	2.50	205
21	16	0.06	17.7	1.75	145	2.25	205	2.75	255
24	14	0.08	25.2	2.00	180	3.00	264	3.00	320
30	14	0.08	30.9	2.50	300	3.50	384	3.50	470
36	12	0.11	51.0	2.75	385	4.00	524	4.00	600
42	12	0.11	59.5			4.50	686	4.50	750
48	12	0.11	68.0			5.00	866	5.00	1000
54	12	0.11	77.8			5.50	1070	5.50	1250
60	10	0.14	108.9			6.00	1295	6.00	1500
66	10	0.14	119.4			6.50	1543		
72	10	0.14	130.4			7.00	1810	7.00	
78	8	0.17	170.6			7.50	2250		
84	8	0.17	185.2			8.00	2409	8.00	

PIPE—SOIL

Sizes (In.)	Standard Wt./Jt. (Lbs.)
2	18
3	26
4	35
5	45
6	55
7	75
8	85
10	115
12	165
15	225

All 5-inch joint lengths

PIPE—STANDARD STEEL

Size (In.)	Wt./Ft. (Lbs.)		Test Pressure (Lbs.)	
	Plain Ends	Thds & Cdgs.	Butt	Lap
1/8	.24	.24	700	. . . .
1/4	.424	.425	700	. . . .
3/8	.567	.568	700	. . . .
1/2	.850	.852	700	. . . .
3/4	1.130	1.134	700	. . . .
1	1.678	1.684	700	. . . .
1 1/4	2.272	2.281	700	1000
1 1/2	2.717	2.731	700	1000
2	3.652	3.678	700	1000
2 1/2	5.793	5.819	800	1000
3	7.575	7.616	800	1000
3 1/2	9.109	9.202	. . . .	1000
4	10.790	10.889	. . . .	1000
4 1/2	12.538	12.642	. . . .	1000
5	14.617	14.810	. . . .	1000
6	18.974	19.185	. . . .	1000
7	23.544	23.769	. . . .	1000
8	24.696	25.000	. . . .	800
8	28.554	28.809	. . . .	1000
9	33.907	34.188	. . . .	900
10	31.201	32.000	. . . .	600
10	34.240	35.000	. . . .	800
10	40.483	41.132	. . . .	900
11	45.557	46.247	. . . .	800
12	43.773	45.000	. . . .	600
12	49.562	50.706	. . . .	800

PLASTER—AGGREGATE PROPORTIONS, COVERAGE

Number of Plaster Coats	PLASTER TO SAND BY WEIGHT				VERMICULITE-PERLITE CUBIC FEET/100 POUNDS OF PLASTER			
	Gypsum Lath	Metal Lath	Insulation Lath	Clay Tile Brick Gypsum Tile	Gypsum Lath	Metal Lath	Insulation Lath	Clay Tile Brick Gypsum Tile
Three Coat Work:								
Scratch Coat Mortar	1:2	1:2	1:2	1:3	1:2	1:2	1:2	1:3
Brown Coat Mortar	1:3	1:3	1:3	1:3	1:3	1:3	1:3	1:3
Two Coat Work:								
Double Back Plastering	1:2 ½	---	---	1:3	1:2 ½	---	---	1:3
Sq. Yds. Coverage/ Ton Plaster	165 to 225	75 to 125	165 to 225	165 to 250	165 to 225	75 to 125	165 to 225	165 to 250

PLASTER—CONCRETE BOND

- Weight—100 lbs./sack
- Coverage—85 – 125 sq. yards/sack, 3/8 inch thick

PLASTER—FINISH (QUANTITY OF MATERIALS)

	Gauging Plaster to Lime	Troweling Keene's to Lime	
Mix by weight of dry materials	1:2	2:1 Medium Hard	4.1 Hard
Color	White or Gray	White	White
Finish Texture	Smooth	Smooth	Smooth
	1 ton gauging 2 tons lime	2 tons Keene's 1 ton lime	4 tons Keene's 1 ton lime
Coverage	900 – 1500 yds.	900 – 1300 yds.	1200 – 1600 yds.

Plaster—Mix Properties, see page 215.

PLASTER—MIXTURE RATIO
(PLUS SMOOTH WHITE FINISH)

100 Square Yards

Material	For Gypsum Plaster 1:2	For Portland Cement Plaster 1:3
Plastering Sand	10,400 lbs.	13,500 lbs.
Gypsum	4,500 lbs.	
Portland Cement		4,400 lbs.
Hydrated Lime		440 lbs.
Finishing Lime	900 lbs.	900 lbs.
Gauging Plaster	300 lbs.	300 lbs.

PLASTER—MIX PROPERTIES

Plaster	Gypsum Wood Fiber		Gyp. Cement Plaster To Sand by weight			Sanded Gyp. Plaster	Gyp. Plaster and Vermiculite		Gyp. Plaster and Perlite		Concrete Bond Plaster
Mix	Neat	1:1	1:1	1:2	1:3	Mill Mix	100 lbs. 2 cu. ft.	100 lbs. 3 cu. ft.	100 lbs. 2 cu. ft.	100 lbs. 3 cu. ft.	
Compressive strength, (lbs./sq. in.)	2550	2000	2040	1160	750	1000	530	310	1030	625	2200
Tensile strength, (lbs./sq. in.)	450	275	250	200	130	200	135	100	170	100	410

PLASTER—VERMICULITE
3/8-Inch Thick/100 Square Yards

Mix	100 Lb. Bags Plaster	Cu. Ft. Vermiculite
1 to 1 1/2	11	17
1 to 2	10	20
1 to 2 1/2	9	23
1 to 3	8	24
1 to 3 1/2	7	25
1 to 4	6	26

PLASTER—SPECIFICATIONS (SAND, VERMICULITE, PERLITE)

Screen Size	% Retained	
	Max.	Min.
No. 4	0	0
8	10	0
30	80	15
50	95	70
100	—	95

Vermiculite wt. 7 1/2 to 10
Perlite 7 1/2 to 12 lbs./cu. ft.

PLASTERS—PACKAGE SIZES

Type	Pkg. Size (Lbs.)
Acoustical	50
Dental	100
Industrial Moulding	100
Moulding	100
Neat Cement	100
Patching	25-11-5
Perlite (Plant mixed)	80
Plaster of Paris	100
Wood Fiber	100-50-25

PLASTIC PANELS—CORRUGATED

Widths Overall (In.)	Thickness (In.)	Wt./ Sq. Ft.	Corrugation (In.)	Static Load/ Sq. Ft. (Lbs.)	Total Thickness (In.)	Pitch (In.)	Width Coverage (In.)
26	$1/16$	7 oz.	1 $1/4$	180	$1/4$	1.26	24
26	$1/16$	7 oz.	2 $1/2$	180	$1/2$	2.67	24
33 $3/4$	$1/16$	7 oz.	2.67	180	$7/8$	2.67	31
42	$1/16$	7 oz.	4.2	180	1 $1/2$	4.2	42

Several types available—product specifications vary

Plate—Steel (Weights), see page 217.

PLATE—STEEL SAFETY (SIZES)

Lbs./ Sq.ft.	Gauge (In.)	Lengths (In.)				
		Widths				
		24	36	48	60	72
8.70	$3/16$	360	360	360	360	360
11.25	$1/4$	360	360	360	360	360
13.80	$5/16$	360	360	360	360	360
16.35	$3/8$	360	360	360	360	360
21.45	$1/2$	360	360	360	360	360
26.55	$5/8$	360	360	360	360	360
3.00	16	192	192	192	. . .	. . .
3.75	14	192	192	192	. . .	. . .
4.50	13	240	240	240	. . .	. . .
5.25	12	240	240	240	240	. . .
6.15	$1/8$	240	240	240	240	240

PLATE—STEEL (WEIGHTS)

Thickness (In.)	Wt. (Lbs.)/ Sq. In.	Wt. (Lbs.)/ Sq. Ft.
3/16	.054	8
1/4	.071	10
5/16	.089	13
3/8	.106	15
7/16	.124	18
1/2	.142	20
9/16	.159	23
5/8	.177	26
11/16	.195	28
3/4	.213	31
13/16	.230	33
7/8	.248	36
1	.283	41
1 1/8	.319	46
1 1/4	.354	51
1 3/8	.390	56
1 1/2	.425	61
1 5/8	.460	66
1 3/4	.496	71
1 7/8	.531	77
2	.567	82
2 1/8	.602	87
2 1/4	.638	92
2 1/2	.708	102
2 3/4	.779	112
3	.850	122
3 1/4	.921	133
3 1/2	.992	143
3 3/4	1.063	153
4	1.133	163
4 1/4	1.204	173
4 1/2	1.275	184
5	1.417	204
5 1/2	1.558	224
6	1.700	245
6 1/2	1.842	265
7	1.983	286
7 1/2	2.125	306
8	2.267	326
9	2.550	367
10	2.833	408

PLYWOOD—HARDWOOD & DECORATIVE

Categories of Commonly Used Species Based on Specific Gravity Ranges[a]

Category A species (0.56 or more specific gravity)	Category B species (0.43 through 0.55 specific gravity)	Category C species (0.42 or less specific gravity)
Ash, Commercial White	Ash, Black	Alder, Red
Beech, American	Avodire	Aspen
Birch, Yellow Sweet	Bay	Basswood, American
Bubinga	Cedar, Eastern Red[b]	Box Elder
Elm, Rock	Cherry, Black	Cativo
Madrone, Pacific	Chestnut, American	Cedar, Western Red[b]
Maple,Black (hard)	Cypress[b]	Ceiba
Maple Sugar (hard)	Elm, American (white, red, or gray)	Cottonwood, Black
Oak, Commercial Red	Fir, Douglas[b]	Cottonwood, Eastern
Oak, Commercial White	Gum, Black	Pine, White and Ponderosa[b]
Oak, Oregon	Gum, Sweet	Poplar, Yellow
Paldao	Hackberry	Redwood[b]
Pecan, Commercial	Lauan, (Philippine Mahogany)	Willow, Black
Rosewood	Limba	
Sapele	Magnolia	
Teak	Mahogany, African	
	Mahogany, Honduras	
	Maple, Red (soft)	
	Maple, Silver (soft)	
	Prima Vera	
	Sycamore	
	Tupelo, Water	
	Walnut, American	

(a) Based on oven-dry weight and volume at 12 percent moisture content.
(b) Softwood.

PLYWOOD—HARDWOOD & DECORATIVE

Types: Technical (Exterior); Type I (Exterior); Type II (Interior); Type III (Interior)

Cores	Plies
Hardwood veneer	3-ply and up
Softwood veneer	3-ply and up
Hardwood lumber	3-, 5-, 7-ply
Softwood lumber	3-, 5-, 7-ply
Particleboard	3-, 5-ply
Hardboard	3-ply
Special	3-ply and up

- Grades: Premium (A); Good (1); Sound (2); Utility (3); Backing (4); Specialty (SP).
- Sizes: Most common—48 by 84 inches, 48 by 96 inches, 48 by 120 inches; other available.
- Thicknesses; Usual range—$1/8$ to $3/4$ inch.
- Identification: By species name of face veneer, which may be either a hardwood or a softwood.

PLYWOOD—SOFTWOOD
(APA PANELS FOR CLOSED SOFFITS)

Long Dimension across Supports

Maximum Span (In.) All Edges Supported	Nominal Panel Thickness	Species Group	Nail Size and Type[a]
24	$11/32''$ APA[b]	All Species Groups	6d nonstaining box or casing
32	$15/32''$ APA[b]		6d nonstaining box or casing
48	$19/32''$ APA[b]		8d nonstaining box or casing

(a) Space nails 6 inches at panel edges and 12 inches at intermediate supports for spans less than 48 inches; 6 inches at all supports for 48-inch spans.

(b) Any suitable grade of Exterior panel which meets appearance requirements.

PLYWOOD—SOFTWOOD (APA GLUED FLOOR SYSTEM)

Maximum Joist Spans with APA RATED STURD-I-FLOOR Panels

Species and Grade of Joist	Joist Size	Joists @ 16" oc		Joists @ 19.2" oc		Joists @ 24" oc
		Sturd-I-Floor 16 or 20 oc	Sturd-I-Floor 24 oc	Sturd-I-Floor 20 oc	Sturd-I-Floor 24 oc	Sturd-I-Floor 24 oc
Douglas fir-Larch No. 1	2×6	11-0	11-4	10-6	10-6	9-5
	2×8	14-3	14-7	13-7	13-10	12-5
	2×10	17-11	18-3	17-0	17-4	15-10
	2×12	21-7	21-11	20-6	20-10	19-3
Douglas fir-Larch No. 2	2×6	10-6	10-6	9-7	9-7	8-7
	2×8	13-10	13-10	12-7	12-7	11-3
	2×10	17-7	17-7	16-1	16-1	14-5
	2×12	21-3	21-5	19-7	19-7	17-6
Douglas fir-South No. 1	2×6	10-4	10-8	9-11	10-2	9-1
	2×8	13-4	13-8	12-8	13-0	12-0
	2×10	16-8	17-0	15-11	16-3	15-4
	2×12	20-1	20-5	19-1	19-5	18-4
Douglas Fir-South No. 2	2×6	10-1	10-1	9-3	9-3	8-3
	2×8	13-1	13-4	12-2	12-2	10-11
	2×10	16-4	16-8	15-6	15-6	13-11
	2×12	19-8	20-0	18-8	18-10	16-11
Hem-Fir No. 1	2×6	10-3	10-3	9-5	9-5	8-5
	2×8	13-7	13-7	12-5	12-5	11-1
	2×10	17-0	17-4	15-10	15-10	14-2
	2×12	20-6	20-10	19-3	19-3	17-2

Species/Grade	Size					
Hem-Fir No.2	2×6	9-4	9-4	8-6	8-6	7-7
	2×8	12-4	12-4	11-3	11-3	10-0
	2×10	15-8	15-8	14-4	14-4	12-10
	2×12	19-1	19-1	17-5	17-5	15-7
Southern Pine KD15 No. 1	2×6	11-0	11-4	10-6	10-10	9-8
	2×8	14-3	14-7	13-7	13-11	12-9
	2×10	17-11	18-3	17-0	17-4	16-3
	2×12	21-7	21-11	20-6	20-10	19-7
Southern Pine KD15 No. 2	2×6	10-8	10-8	9-9	9-9	8-8
	2×8	13-10	14-0	12-10	12-10	11-6
	2×10	17-4	17-8	16-4	16-4	14-8
	2×12	20-11	21-2	19-10	19-11	17-9
Southern Pine MC19 No. 1	2×6	10-10	11-2	10-4	10-4	9-3
	2×8	14-0	14-4	13-4	13-8	12-2
	2×10	17-8	17-11	16-9	17-1	15-7
	2×12	21-3	21-7	20-2	20-6	18-11
Southern Pine MC19 No. 2	2×6	10-3	10-3	9-5	9-5	8-5
	2×8	13-7	13-7	12-5	12-5	11-1
	2×10	17-4	17-4	15-10	15-10	14-2
	2×12	20-11	21-1	19-3	19-3	17-2

(a) Based on live load of 40 psf, total load of 50 psf, deflection limited to 1/360 at 40 psf.

(b) Glue tongue-and-groove joints. If square-edge panels are used, block panel edges and glue between panels and between panels and blocking.

(c) Contact truss manufacturer for floor truss or plywood "I" joist span recommendations.

PLYWOOD—SOFTWOOD
(APA CLASSIFICATION OF SPECIES)

Group 1	Group 2		Group 3	Group 4	Group 5

Apitong
Beech,
  American
Birch
  Sweet
  Yellow
Douglas
  Fir 1 (1)
Kapur
Keruing
Larch,
  Western
Maple, Sugar
Pine
  Caribbean
  Ocote
Pine, Southern
  Loblolly
  Longleaf
  Shortleaf
  Slash
Tanoak
White
	Cedar, Port
	  Orford
	Cypress
	Douglas
	  Fir 2 (1)
	Fir
	  Balsam
	  California
	    Red
	  Grand
	  Noble
	  Pacific
	    Silver
	  White
	Hemlock,
	  Western
	Lauan
	  Almon
	  Bagtikan
	  Mayapis
	  Red
	  Tangile
		Maple, Black
		Mengkulang
		Meranti,
		  Red (2)
		Mersawa
		Pine
		  Pond
		  Red
		  Virginia
		  Western
		    White
		Spruce
		  Black
		  Red
		  Sitka
		Sweetgum
		Tamarack
		Yellow-
		  Poplar
			Alder, Red
			Birch, Paper
			Cedar, Alaska
			Fir,
			  Subalpine
			Hemlock,
			  Eastern
			Maple,
			  Bigleaf
			Pine
			  Jack
			  Lodgepole
			  Ponderosa
			  Spruce
			Redwood
			Spruce
			  Engelmann
			  White
				Aspen
				  Bigtooth
				  Quaking
				Cativo
				Cedar
				  Incense
				  Western
				
				Cottonwood
				  Eastern
				  Black
				  (Western
				  Poplar)
				Pine
				  Eastern
				    White
				  Sugar
					Basswood
					Poplar,
					  Balsam

(1) Douglas Fir from trees grown in the states of Washington, Oregon, California, Idaho, Montana, Wyoming, and the Canadian Provinces of Alberta and British Columbia shall be classed as Douglas Fir No. 1. Douglas Fir from trees grown in the states of Nevada, Utah, Colorado, Arizona and New Mexico shall be classed as Douglas Fir No. 2.

(2) Red Meranti shall be limited to species having a specific gravity of 0.41 or more based on green volume and oven-dry weight.

PLYWOOD—SOFTWOOD (APA PANEL WALL SHEATHING[a])
Panels Continuous Over Two or More Spans.

Panel Span Rating	Maximum Stud Spacing (In.)	Nail Size[b]	Nail Spacing (In.)[b]	
			Supported Panel Edges	Intermediate Supports
12/0, 16/0, 20/0 or Wall-16 oc	16	6d for panels 1/2" thick or less; 8d for thicker panels	6	12
24/0, 24/16, 32/16 or Wall-24 oc	24			

(a) See requirements for nailable panel sheathing when exterior covering is to be nailed to sheathing.
(b) Common smooth, annular, spiral-thread, or galvanized box.
(c) Other code-approved fasteners may be used.

PLYWOOD—SOFTWOOD (APA PANELS FOR OPEN SOFFIT OR FOR COMBINED ROOF DECKING-CEILING[a][b])

Long Dimension across Supports

Maximum Span (In.)	Panel Description (All Panels Exterior, Exposure 1 or Interior with Exterior Glue)	Species Group for Plywood
16	15/32" APA RATED SIDING 303 15/32" APA MDO and Sanded Plywood APA RATED STURD-I-FLOOR 16 oc	1,2,3,4 1,2,3,4 —
24	15/32" APA RATED SIDING 303 15/32" APA MDO and Sanded Plywood 19/32" APA RATED SIDING 303 19/32" APA MDO and Sanded Plywood APA RATED STURD-I-FLOOR 16 oc	1 1,2,3 1,2,3,4 1,2,3,4 —
32	19/32" APA RATED SIDING 303 19/32"APA MDO and Sanded Plywood 23/32" APA Textured Plywood 23/32" APA MDO and Sanded Plywood APA RATED STURD-I-FLOOR 20 oc	1 1 1,2,3,4 1,2,3,4 —
48	1 1/8" APA Textured Plywood APA RATED STURD-I-FLOOR 48 oc	1,2,3,4 —

(a) All panels will support at least 30 psf live load plus 10 psf dead loaf at maximum span.
(b) For appearance purposes, blocking, tongue-and-groove edges or other suitable edge supports should be provided.

PLYWOOD—SOFTWOOD (APA PANEL SUBFLOORING[a])

APA Rated Sheathing

Panel Span Rating (or Group Number)	Panel Thickness (In.)	Maximum Span (In.)	Nail Size & Type[g]	Nail Spacing (In.)	
				Supported Panel Edges	Intermediate Supports[f]
24/16	7/16	16	6d common	6	12
32/16	15/32, 1/2, 5/8	16[b]	8d common[c]	6	12
40/20	9/16, 19/32, 5/8, 3/4, 7/8	20[d]	8d common	6	12
48/24	23/32, 3/4, 7/8	24	8d common	6	12
1 1/18" Groups 1 & 2[e]	1 1/8	48	10d common	6	6

(a) For subfloor recommendations under gypsum concrete, contact manufacturer of floor topping.
(b) Span may be 24 inches if 3/4-inch wood strip flooring is installed at right angles to joists.
(c) 6d common nail permitted if panel is 1/2 inch or thinner.
(d) Span may be 24 inches if 3/4-inch wood strip flooring is installed at right angles to joists, or if a minimum of 1 1/2 inches of lightweight concrete is applied over panels.
(e) Check dealer for availability.
(f) Applicable building codes may require 10" oc nail spacing at intermediate supports for floors.
(g) Other code-approved fasteners may be used.

PLYWOOD—SOFTWOOD
(APA PLYWOOD SYSTEMS FOR CERAMIC TILE FLOORING)

Based on ANSI Standard A108 and Recommendations of the Tile Council of America.

Joist Spacing (In.)	Minimum Panel Thickness (In.)		Tile Installation
	Subfloor[a]	Underlayment[b]	
Residential 16	$15/32$	(d)	"Dry-Set" mortar; or latex - Portland Cement mortar
16	$19/32$	—	Cement mortar ($3/4"$ - $1 1/4"$)
16	$19/32$	$11/32$	Organic adhesive
16	$19/32$	$15/32$[e]	Epoxy mortar
16	$19/32$ T&G[c][e]	—	Epoxy mortar
Commercial 16	$15/32$	(d)	"Dry-Set" mortar; or latex - Portland Cement mortar
16	$19/32$	—	Cement mortar ($3/4"$ - $1 1/4"$)
16	$19/32$	$19/32$[c][e]	Epoxy mortar

(a) APA RATED SHEATHING with subfloor Span Rating of 16″ oc ($15/32$″ panel) or 20″ oc ($19/32$″ panel), except as noted

(b) APA Underlayment or sanded Exterior grade except as noted

(c) APA RATED STURD-I-FLOOR with 20″ oc Span Rating

(d) Bond glass mesh mortar units to subfloor with latex - Portland Cement mortar prior to spreading mortar for setting ceramic tile

(e) Leave $1/4$″ space at panel ends and edges: trim panels as necessary to maintain end spacing and panel support on framing. Fill joints with epoxy mortar when it is spread for setting tile. With single-layer residential floors, use solid lumber blocking or framing under all panel and edge joints (including T&G joints)

PLYWOOD—SOFTWOOD (APA PLYWOOD UNDERLAYMENT[d])

Plywood Grades[a]	Application	Minimum Plywood Thickness (In.)	Fastener Size and Type[e]	Fastener Spacing (In.)[b]	
				Panel Edges	Intermediate
APA UNDERLAYMENT APA C–C Plugged EXT APA RATED STURD-I-FLOOR (¹⁹/₃₂" or thicker)	Over smooth subfloor	¹/₄	3d ring-shank nails[c]	3	6 each way
	Over lumber subfloor or other uneven surfaces	¹¹/₃₂	3d ring-shank nails[c]	6	8 each way
Same grades as above, but species Group 1 only.	Over lumber floor up to 4" wide. Face grain must be perpendicular to boards.	¹/₄	3d ring-shank nails[c]	3	6 each way

(a) In areas to be finished with thin floor coverings such as tile or sheet vinyl, specify Underlayment. C – C Plugged or STURD-I-FLOOR with "sanded face." Underlayment A – C. Underlayment B – C. Marine EXT or sanded plywood grades marked "Plugged Crossbands Under Face." "Plugged Crossbands (or Core)," "Plugged Inner Plies" or "Meets Underlayment Requirements" may also be used under thin floor coverings.

(b) Space fasteners so they do not penetrate framing.

(c) Use 3d ring-shank nails for ¹/₂ inch panels and 4d ring-shank nails for ⁵/₈ inch or ³/₄ inch panels.

(d) For underlayment recommendations under ceramic tile refer to pg. 225.

(e) Other code-approved fasteners may be used.

PLYWOOD—SOFTWOOD (APA RATED SIDING)
Over Nailable Sheathing

Siding Description[a]		Nominal Thickness (In.) or Span Rating	Maximum Spacing of Vertical Rows of Nails (In.)		Nail Size Box, Siding or Casing Nails[b]	Nail Spacing[c] (In.)	
			Long Dimension Vertical	Long Dimension Horizontal		Panel Edges	Intermediate
Panel Siding	APA MDO EXT	11/32 & 3/8	16	24	6d for panels 1/2" thick or less; 8d for thicker panels	6	12
	APA RATED SIDING EXT	1/2 & thicker	24	24			
		16 oc (including T1-11) 24 oc	16	24			
			24	24			
Lap Siding	APA MDO EXT	11/32 & thicker	—	—	6d for siding 1/2" thick or less; 8d for thicker siding	8 along bottom edge	—
	APA RATED SIDING—LAP EXT	1/2 & thicker, or 16 oc or 24 oc	—	—			

(a) Recommendations apply to all species groups for veneered APA Rated Siding, including APA 303 Siding.
(b) Hot-dipped or hot-tumbled galvanized steel nails are recommended for most siding applications. For best performance, stainless steel nails or aluminum nails should be considered. APA tests also show that electrically or mechanically galvanized steel nails appear satisfactory when plating meets or exceeds thickness requirements of ASTM A641 Class 2 coatings, and is further protected by yellow chromate coating.
Note: Galvanized fasteners may react under wet conditions with the natural extractives of some wood species and may cause staining if left unfinished. Such staining can be minimized if the siding is finished in accordance with APA recommendations, or if the roof overhang protects the siding from direct exposure to moisture and weathering.
(c) Recommendations of siding manufacturer may vary.

PLYWOOD—SOFTWOOD (APA RATED STURD-I-FLOOR)[a]

Span Rating (Maximum Joist Spacing) (In.)	Panel Thickness[a] (In.)	Fastening: Glue-Nailed[c]			Fastening: Nailed Only		
		Nail Size and Type	Spacing (In.)		Nail Size and Type	Spacing (In.)	
			Supported Panel Edges	Intermediate Supports		Supported Panel Edges	Intermediate Supports[g]
16	19/32, 5/8, 21/32	6d ring- or screw-shank[d]	12	12	6d ring- or screw-shank	6	12
20	19/32, 5/8, 23/32, 3/4	6d ring- or screw-shank[d]	12	12	6d ring- or screw-shank	6	12
24	11/16, 23/32, 3/4	6d ring- or screw-shank[d]	12	12	6d ring- or screw-shank	6	12
	7/8, 1	8d ring- or screw-shank[d]	6	12	8d ring- or screw-shank	6	12
32	7/8, 1	8d ring- or screw-shank[d]	6	12	8d ring- or screw-shank	6	12
48	1 1/8	8d ring- or screw-shank[e]	6	(f)	8d ring- or screw-shank[e]	6	(f)

(a) Special conditions may impose heavy traffic and concentrated loads that require construction in excess of the minimums shown. See page 80 for heavy duty floor recommendations.

(b) As indicated above, panels in a given thickness may be manufactured in more than one Span Rating. Panels with a Span Rating greater than the actual joist spacing may be substituted for panels of the same thickness with a Span Rating matching the actual joist spacing. For example, 19/32-inch-thick Sturd-I-Floor 20 oc may be substituted for 19/32 -inch-thick Sturd-I-Floor 16 oc over joists 16 inches on center.

(c) Use only adhesives conforming to APA Specification AFG-01, applied in accordance with the manufacturer's recommendations. If nonveneered panels with sealed surfaces and edges are to be used, use only solvent-based glues; check with panel manufacturer.

(d) 8d common nails may be substituted if ring- or screw-shank nails are not available.

(e) 10d common nails may be substituted with 1 1/8-inch panels if supports are well seasoned.

(f) Space nails 6 inches for 48-inch spans and 12 inches for 32-inch spans.

(g) Applicable building codes may require 10" oc nail spacing at intermediate supports for floors.

PLYWOOD—SOFTWOOD (APA STURD-I-WALL CONSTRUCTION)

Recommendations Apply to APA RATED SIDING Direct to Studs and Over Nonstructural Sheathing.

Siding Description[a]		Nominal Thickness (In.) or Span Rating	Max. Stud Spacing (In.)		Nail Size (Use nonstaining Box, Siding, or Casing Nails) [b][c]	Nail Spacing[e] (In.)	
			Long Dimension Vertical	Long Dimension Horizontal		Panel Edges	Intermediate Supports
Panel Siding	APA MDO EXT	11/32 & 3/8	16	24	6d for siding 1/2" thick or less; 8d for thicker siding	6(d)	12
		1/2 & thicker	24	24			
	APA RATED SIDING EXT	16 oc (including T1-11)	16	24			
		24 oc	24	24			
Lap Siding	APA RATED SIDING—LAP EXT	16 oc	—	16	6d for siding 1/2" thick or less; 8d for thicker siding	16 along bottom edge	—
		24 oc	—	24		24 along bottom edge	—

(a) Recommendations apply to all species groups for veneered APA Rated Siding, including APA 303 Siding.

(b) If panel siding is applied over sheathing thicker than 1/2 inch, use next regular nail size. If lap siding is installed over rigid foam insulation sheathing up to 1 inch thick, use 10d (3") nails for 3/8" or 7/16" siding, 12d (3 1/4") nails for 15/32" or 1/2" siding, and 16d (3 1/2") nails for 19/32" or thicker siding. Use nonstaining box nails for siding installed over foam insulation sheathing.

(c) Hot-dipped or hot-tumbled galvanized steel nails are recommended for most siding applications. For best performance, stainless steel nails or aluminum nails should be considered. APA tests also show that electrically or mechanically galvanized steel nails appear satisfactory when plating meets or exceeds thickness requirements of ASTM A641 Class 2 coatings.and is further protected by yellow chromate coating.

Note: Galvanized fasteners may react under wet conditions with the natural extractives of some wood species and may cause staining if left unfinished. Such staining can be minimized if the siding is finished in accordance with APA recommendations, or if the roof overhang protects the siding from direct exposure to moisture and weathering.

(d) For braced wall section with 11/32" or 3/8" panel siding applied horizontally over studs 24" oc, space nails 3" oc along panel edges.

(e) Recommendations of siding manufacturer may vary.

PLYWOOD-SOFTWOOD (APA VENEER GRADES)

N	Smooth surface "natural finish" veneer. Select, all heartwood or all sapwood. Free of open defects. Allows not more than 6 repairs, wood only, per 4 × 8 panel, made parallel to grain and well matched for grain and color.
A	Smooth, paintable. Not more than 18 neatly made repairs, boat, sled, or router type, and parallel to grain, permitted. May be used for natural finish in less demanding applications. Synthetic repairs permitted.
B	Solid surface. Shims, circular repair plugs and tight knots to 1 inch across grain permitted. Some minor splits permitted. Synthetic repairs permitted.
C **Plugged**	Improved C veneer with splits limited to $1/8$-inch width and knotholes and borer holes limited to $1/4 \times 1/2$ inch. Admits some broken grain. Synthetic repairs permitted.
C	Tight knots to $1 1/2$ inch. Knotholes to 1 inch across grain and some to $1 1/2$ inch if total width of knots and knotholes is within specified limits. Synthetic or wood repairs. Discoloration and sanding defects that do not impair strength permitted. Limited splits allowed. Stitching permitted.
D	Knots and knotholes to $2 1/2$-inch width across grain and $1/2$ inch larger within specified limits. Limited splits are permitted. Stitching permitted. Limited to Exposure 1, or Interior panels.

PLYWOOD—SOFTWOOD

GUIDE TO APA SANDED &
TOUCH-SANDED PLYWOOD
Trademarks Shown
Are Typical Facsimiles

| A·A·G·1·EXPOSURE1·APA·000·PS1-83 |

APA A-A

Use where appearance of both sides is important for interior applications such as built-ins, cabinets, furniture, partitions; and exterior applications such as fences, signs, boats, shipping containers, tanks, ducts, etc. Smooth surfaces suitable for painting. EXPOSURE DURABILITY CLASSIFICATIONS: Interior, Exposure 1, Exterior. COMMON THICKNESSES: $1/4$, $11/32$, $3/8$, $15/32$, $1/2$, $19/32$, $5/8$, $23/32$, $3/4$.

| A·B·G·1·EXPOSURE1·APA·000·PS1-83 |

APA A-B

For use where appearance of one side is less important but where two solid surfaces are necessary. EXPOSURE DURABILITY CLASSIFICATIONS: Interior, Exposure 1, Exterior. COMMON THICKNESSES: $1/4$, $11/32$, $3/8$, $15/32$, $1/2$, $19/32$, $5/8$, $23/32$, $3/4$.

APA A-C GROUP 1 EXTERIOR 000 PS 1-83

APA A-C

For use where appearance of one side is important in exterior applications such as soffits, fences, structural uses, boxcar and truck linings, farm buildings, tanks, trays, commercial refrigerators, etc.[1] EXPOSURE DURABILITY CLASSIFICATION: Exterior. COMMON THICKNESSES: $1/4$, $9/32$, $11/32$, $3/8$, $15/32$, $1/2$, $19/32$, $5/8$, $23/32$, $3/4$.

APA A-D GROUP 1 EXPOSURE 1 000 PS 1-83

APA A-D

For use where appearance of only one side is important in interior applications, such as paneling, built-ins, shelving, partitions, flow racks, etc. [1] EXPOSURE DURABILITY CLASSIFICATIONS: Interior, Exposure 1. COMMON THICKNESSES: $1/4$, $9/32$, $11/32$, $3/8$, $15/32$, $1/2$, $19/32$, $5/8$, $23/32$, $3/4$.

| B·B·G·2·EXPOSURE1·APA·000·PS1-83 |

APA A-B

Utility panels with two solid sides. EXPOSURE DURABILITY CLASSIFICATIONS: Interior, Exposure 1, Exterior. COMMON THICKNESSES: $1/4$, $11/32$, $3/8$, $15/32$, $1/2$, $19/32$, $5/8$, $23/32$, $3/4$.

PLYWOOD—SOFTWOOD

GUIDE TO APA
SPECIALTY PANELS
Trademarks Shown
Are Typical Facsimiles

HDO · A-A · G-1 · EXT-APA · 000 · PS1-83

APA MEDIUM DENSITY OVERLAY (MDO)

Plywood panel manufactured with smooth, opaque, resin-treated fiber overlay providing ideal base for paint on one or both sides. Excellent material choice for shelving, factory work surfaces, paneling, built-ins, signs and numerous other construction and industrial applications. Also available as Rated Siding 303 with texture-embossed or smooth surface on one side only and in Structural I. EXPOSURE DURABILITY CLASSIFICATION: Exterior. COMMON THICKNESSES: $11/32$, $3/8$, $15/32$, $1/2$, $19/32$, $5/8$, $23/32$, $3/4$.

```
——APA——
M. D. OVERLAY
GROUP 1
EXTERIOR
—000—
PS 1-83
```

APA MEDIUM DENSITY OVERLAY (MDO)

Plywood panel manufactured with smooth, opaque, resin-treated fiber overlay providing ideal base for paint on one or both sides. Excellent material choice for shelving, factory work surfaces, paneling, built-ins, signs and numerous other construction and industrial applications. Also available as Rated siding 303 with texture-embossed or smooth surface on one side only and in Structural I. EXPOSURE DURABILITY CLASSIFICATION: Exterior. COMMON THICKNESSES: $11/32$, $3/8$, $15/32$, $1/2$, $19/32$, $5/8$, $23/32$, $3/4$.

MARINE · A-A · EXT-APA · 000 · PS1-83

APA MARINE

Specially designed plywood panel made only with Douglas fir or western larch, solid jointed cores, and highly restrictive limitations on core gaps and face repairs. Ideal for boat hulls and other marine applications. Also available with HDO or MDO faces. EXPOSURE DURABILITY CLASSIFICATION: Exterior. COMMON THICKNESSES: $1/4$, $3/8$, $1/2$, $5/8$, $3/4$.

```
——APA——
DECORATIVE
GROUP 2
INTERIOR
—000—
PS 1-83
```

APA DECORATIVE

Rough sawn, brushed, grooved, or other faces. For paneling, interior accent walls, built-ins, counter facing, exhibit displays, etc. Made by some manufacturers in Exterior for exterior siding, gable ends, fences and other exterior applications. Use recommendations for Exterior panels vary; check with the manufacturer. EXPOSURE DURABILITY CLASSIFICATIONS: Interior, Exposure 1, Exterior. COMMON THICKNESSES: $5/16$, $3/8$, $1/2$, $5/8$.

PLYRON·EXPOSURE1-APA·000

APA PLYRON

APA proprietary plywood panel with hardboard face on both sides. Faces tempered, untempered, smooth or screened. For countertops, shelving, cabinet doors, concentrated load flooring, etc. EXPOSURE DURABILITY CLASSIFICATIONS: Interior, Exposure 1, Exterior. COMMON THICKNESSES: $1/2$, $5/8$, $3/4$.

APA
PLYFORM
B-B CLASS I
EXTERIOR
000
PS 1-83

APA B-B PLYFORM CLASS I

APA proprietary concrete form panels designed for high reuse. Sanded both sides and mill-oiled unless otherwise specified. Class I, the strongest, stiffest and more commonly available, is limited to Group 1 faces, Group 1 or 2 crossbands, and Group 1, 2, 3, or 4 inner plies. (Plyform Class II, limited to Group 1, 2, or 3 faces under certain conditions and Group 1, 2, 3, or 4 inner plies may also be available.) Also available in HDO for very smooth concrete finish, in Structural I, and with special overlays. EXPOSURE DURABILITY CLASSIFICATION: Exterior. COMMON THICKNESSES: $19/32$, $5/8$, $23/32$, $3/4$.

APA
B-C GROUP 1
EXTERIOR
000
PS 1-83

APA B-C

Utility panel for farm service and work buildings, boxcar and truck linings, containers, tanks, agricultural equipment, as a base for exterior coatings and other exterior uses.[1] EXPOSURE DURABILITY CLASSIFICATION: Exterior. COMMON THICKNESSES: $1/4$, $9/32$, $11/32$, $3/8$, $15/32$, $1/2$, $19/32$, $5/8$, $23/32$, $3/4$.

APA
B-D GROUP 2
INTERIOR
000
PS 1-83

APA B-D

Utility panel for backing, sides of built-ins, industry shelving, slip sheets, separator boards, bins and other interior or protected applications.[1] EXPOSURE DURABILITY CLASSIFICATIONS: Interior, Exposure 1. COMMON THICKNESSES: $1/4$, $9/32$, $11/32$, $3/8$, $15/32$, $1/2$, $19/32$, $5/8$, $23/32$, $3/4$.

PLYWOOD—SOFTWOOD

GUIDE TO APA PERFORMANCE-RATED PANELS
Trademarks Shown
Are Typical Facsimiles

```
╔══════════════════╗
    APA
RATED SHEATHING
24/16 7/16 INCH
SIZED FOR SPACING
EXPOSURE 1
    000
NER-QA397  PRP-108
╚══════════════════╝
```

APA RATED SHEATHING

Specially designed for subflooring, wall sheathing and roof sheathing, but also used for broad range of other construction, industrial and do-it-yourself applications. Can be manufactured as conventional plywood, as a composite, or a nonveneer panel (waferboard, oriented strand board, structural particleboard). For special engineered applications, veneered panels conforming to PS 1 may be required. SPAN RATINGS: $16/0$, $20/0$, $24/0$, $24/16$, $32/16$, $40/20$, $48/24$. EXPOSURE DURABILITY CLASSIFICATIONS: Exterior, Exposure 1, Exposure 2. COMMON THICKNESSES: $5/16$, $3/8$, $7/16$, $15/32$, $1/2$, $19/32$, $5/8$, $23/32$, $3/4$.

```
╔══════════════════╗
    APA
RATED SHEATHING
STRUCTURAL I
32/16 15/32 INCH
SIZED FOR SPACING
EXPOSURE 1
    000
PS 1-83  C-D
NER-QA397  PRP-108
╚══════════════════╝
```

```
╔══════════════════╗
    APA
RATED SHEATHING
32/16 15/32 INCH
SIZED FOR SPACING
EXPOSURE 1
    000
STRUCTURAL I RATED
DIAPHRAGMS-SHEAR WALLS
PANELIZED ROOFS
NER-QA397  PRP-108
╚══════════════════╝
```

APA STRUCTURAL I RATED SHEATHING

Unsanded grade for use where cross-panel strength and stiffness or shear properties are of maximum importance, such as panelized roofs, diaphragms and shear walls. All plies in Structural I plywood panels are special improved grades and panels marked PS 1 are limited to Group 1 species. (Structural II plywood panels, limited to Group, 1, 2, or 3 species, may also be available. However, application recommendations for Structural II plywood are identical to those for APA RATED SHEATHING plywood marked PS 1.) SPAN RATINGS: $20/0$, $24/0$, $32/16$, $40/20$, $48/24$. EXPOSURE DURABILITY CLASSIFICATIONS: Exterior, Exposure 1. COMMON THICKNESSES: $5/16$, $3/8$, $15/32$, $1/2$, $19/32$, $5/8$, $23/32$, $3/4$.

```
╔══════════════════╗
    APA
RATED STURD-I-FLOOR
20 oc 19/32 INCH
SIZED FOR SPACING
EXPOSURE 1
    000
NER-QA397  PRP-108
╚══════════════════╝
```

APA RATED STURD-I-FLOOR

Specially designed as combination subfloor-underlayment. Provides smooth surface for application of carpet and possesses high concentrated and impact load resistance. Can be manufactured as conventional plywood, as a composite, or as a nonveneer panel (waferboard, oriented strand board, structural particleboard). Available square edge or tongue-and-groove. SPAN RATINGS: 16, 20, 24, 32, 48. EXPOSURE DURABILITY CLASSIFICATIONS: Exterior, Exposure 1, Exposure 2. COMMON THICKNESSES: $19/32$, $5/8$, $23/32$, $3/4$, $1 1/8$.

```
╔══════════════════╗
    APA
RATED SIDING
24 oc 15/32 INCH
SIZED FOR SPACING
EXTERIOR
    000
NER-QA397  PRP-108
╚══════════════════╝
```

```
╔══════════════════╗
    APA
RATED SIDING
303-18-S/W
16 oc 11/32 INCH
        GROUP 1
SIZED FOR SPACING
EXTERIOR
    000
PS 1-83  FHA-UM-64
NER-QA397  PRP-108
╚══════════════════╝
```

APA RATED SIDING

For exterior siding, fencing, etc. Can be manufactured as conventional veneered plywood, as a composite or as a nonveneer siding. Both panel and lap siding available. Special surface treatment such as V-groove, channel groove, deep groove (such as APA Texture 1-11), brushed, rough-sawn and texture-embossed (MDO). Span Rating (stud spacing for siding qualified for APA Sturd-I-Wall applications) and face grade classification (for veneer-faced siding) indicated in trademark. EXPOSURE DURABILITY CLASSIFICATION: Exterior. COMMON THICKNESSES: $11/32$, $3/8$, $15/32$, $1/2$, $19/32$, $5/8$.

≡APA≡
UNDERLAYMENT
GROUP 1
EXPOSURE 1
000
PS 1-83

≡APA≡
C-C PLUGGED
GROUP 2
EXTERIOR
000
PS 1-83

≡APA≡
C-D PLUGGED
GROUP 2
EXPOSURE 1
000
PS 1-83

APA UNDERLAYMENT

For application over structural subfloor. Provides smooth surface for application of carpet and possesses high concentrated and impact load resistance. Touch-sanded. For areas to be covered with resilient non-textile flooring, specify panels with "sanded face."[2] EXPOSURE DURABILITY CLASSIFICATIONS: Interior, Exposure 1. COMMON THICKNESSES: $1/4$, $9/32$, $11/32$, $3/8$, $1/2$, $19/32$, $5/8$, $23/32$, $3/4$.

APA C-C PLUGGED

For use as an underlayment over structural subfloor, refrigerated or controlled atmosphere storage rooms, pallet fruit bins, tanks, boxcar and truck floors and linings, and other exterior applications. Provides smooth surface for application of carpet and possesses high concentrated and impact load resistance. Touch-sanded. For areas to be covered with resilient non-textile flooring, specify panels with "sanded face." EXPOSURE DURABILITY CLASSIFICATION: Exterior. COMMON THICKNESSES: $11/32$, $3/8$, $1/2$, $19/32$, $5/8$, $23/32$, $3/4$.

APA C-D PLUGGED

For open soffits, buit-ins, cable reels, walkways, separator boards and other interior or protected applications. Not a substitute for Underlayment or APA Rated Sturd-I-Floor as it lacks their puncture resistance. Touch-sanded. EXPOSURE DURABILITY CLASSIFICATIONS: Interior, Exposure 1. COMMON THICKNESSES: $3/8$, $1/2$, $19/32$, $5/8$, $23/32$, $3/4$.

PLYWOOD—SOFTWOOD (INTERIOR PANELING)

Plywood Thickness (In.)	Maximum Support Spacing (In.)	Nail Size (Use Casing or Finishing Nails)	Nail Spacing (In.)	
			Panel Edges	Intermediate Supports
1/4	16[a]	4d	6	12
5/16	16[b]	6d	6	12
3/8, 11/32	24	6d	6	12
1/2, 15/32	24	6d	6	12
5/8, 19/32	24	8d	6	12
3/4	24	8d	6	12
Texture 1-11	24	8d	6	12

(a) Can be 20 inches if face grain of paneling is across supports.
(b) Can be 24 inches if face grain of paneling is across supports.

PLYWOOD—SOFTWOOD (PANEL RECOMMENDATIONS FOR APA GLUED FLOOR SYSTEM)[a]

Joist Spacing (In.)	Flooring Type	APA Panel Grade and Span Rating	Possible Thickness (In.)
16	Carpet and Pad	STURD-I-FLOOR 16 oc, 20 oc	$19/32$, $5/8$, $21/32$
16	Separate Underlayment or Structural Finish Flooring	RATED SHEATHING $24/16$, $32/16$ $40/20$, $48/24$	$7/16$, $15/32$, $1/2$ $19/32$, $5/8$, $23/32$, $3/4$
19.2	Carpet and Pad	STURD-I-FLOOR 20 oc, 24 oc	$19/32$, $5/8$, $23/32$, $3/4$
19.2	Separate Underlayment or Structural Finish Flooring	RATED SHEATHING $40/20$, $48/24$	$19/32$, $5/8$, $23/32$, $3/4$
24	Carpet and Pad	STURD-I-FLOOR 24 oc, 32 oc	$11/16$, $23/32$, $3/4$, $7/8$, 1
24	Separate Underlayment or Structural Finish Flooring	RATED SHEATHING $48/24$	$23/32$, $3/4$
32	Carpet and Pad	STURD-I-FLOOR 32 oc, 48 oc	$7/8$, 1, 1 $1/8$
48	Carpet and Pad	STURD-I-FLOOR 48 oc	1 $3/32$, 1 $1/8$

(a) For panel recommendations under ceramic tile, refer to on pg. 225.

PLYWOOD—SOFTWOOD (PS 1 PLYWOOD RECOMMENDATIONS FOR UNIFORMLY LOADED HEAVY DUTY FLOORS(a))

Deflection Limited to 1/240 of Span.

Span Ratings Apply to APA RATED SHEATHING and APA RATED STURD-I-FLOOR, Respectively, Marked PS 1.

Uniform Live Load (psf)	Center-to-Center Support Spacing (Inches) (Nominal 2-Inch-Width Supports Unless Noted)					
	12(b)	16(b)	20(b)	24(b)	32	48(c)
50	32/16, 16 oc	32/16, 16 oc	40/20, 20 oc	48/24, 24 oc	48 oc	48 oc
100	32/16, 16 oc	32/16, 16 oc	40/20, 20 oc	48/24, 24 oc	48 oc	1 1/2(d)
150	32/16, 16 oc	32/16, 16 oc	40/20, 20 oc	48/24, 48 oc	48 oc	1 3/4(e), 2(d)
200	32/16, 16 oc	40/20, 20 oc	48/24, 24 oc	48 oc	1 1/8(e), 1 3/8(d)	2 (e), 2 1/2(d)
250	32/16, 16 oc	40/20, 24 oc	48/24, 48 oc	48 oc	1 3/8(e), 1 1/2(d)	2 1/4(e)
300	32/16, 16 oc	48/24, 24 oc	48 oc	48 oc	1 1/2(e), 1 5/8(d)	2 1/4(e)
350	40/20, 20 oc	48/24, 48 oc	48 oc	1 1/8(e), 1 3/8(d)	1 1/2(e), 2(d)	
400	40/20, 20 oc	48 oc	48 oc	1 1/4(e), 1 3/8(d)	1 5/8(e), 2(d)	
450	40/20, 24 oc	48 oc	48 oc	1 3/8(e), 1 1/2(d)	2(e), 2 1/4(d)	
500	48/24, 24 oc	48 oc	48 oc	1 1/2(d)	2(e), 2 1/4(d)	

(a) Use plywood with T & G edges, or provide structural blocking at panel edges or install a separate underlayment.

(b) A-C Group 1 sanded plywood panels may be substituted for span rated Sturd-I-Floor panels (1/2-inch for 16 oc; 5/8-inch for 20 oc; 3/4-inch for 24 oc).

(c) Nominal 4-inch wide supports.

(d) Group 1 face and back, any species inner plies, sanded or unsanded, single layer.

(e) Structural plywood, sanded or unsanded, single layer.

PLYWOOD—SOFTWOOD (REDWOOD PROFILES)

Saw-textured plywood is available in the following standard sizes and patterns (all available in lengths of 8, 9, and 10 feet). All patterns are available in 3/8-inch thickness on special order, with groove spacings as shown but grooves 1/16 inch deep. Other special patterns are available on quantity orders. Panels are shiplapped on long edges only.

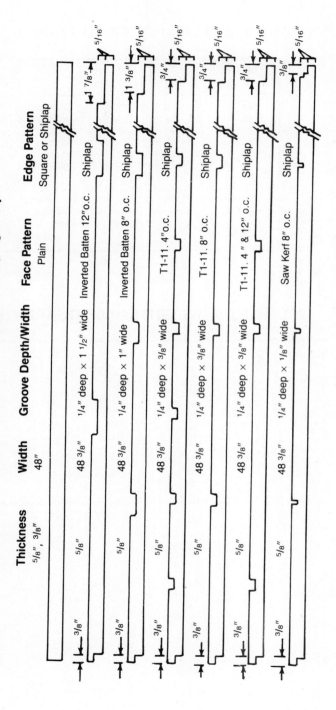

Thickness	Width	Groove Depth/Width	Face Pattern	Edge Pattern
5/8", 3/8"	48"		Plain	Square or Shiplap
5/8"	48 3/8"	1/4" deep × 1 1/2" wide	Inverted Batten 12" o.c.	Shiplap
5/8"	48 3/8"	1/4" deep × 1" wide	Inverted Batten 8" o.c.	Shiplap
5/8"	48 3/8"	1/4" deep × 3/8" wide	T1-11. 4" o.c.	Shiplap
5/8"	48 3/8"	1/4" deep × 3/8" wide	T1-11. 8" o.c.	Shiplap
5/8"	48 3/8"	1/4" deep × 3/8" wide	T1-11. 4" & 12" o.c.	Shiplap
5/8"	48 3/8"	1/4" deep × 1/8" wide	Saw Kerf 8" o.c.	Shiplap

PLYWOOD—SOFTWOOD (USES)

Thickness (In.)	Use
1/4 or 1/2	Interior Wall and Ceiling Coverings
1/4 or 3/8	(Camps, barracks, residences, etc.)
5/16 or 3/8	Wall Sheathing (To be covered)
5/16	Roof Sheathing (To be covered)
3/8	Roof Sheathing (To be covered)
5/8	Roof Sheathing (To be covered)
1/2, 5/8	Sub-Floors
3/8 or more	Exterior Panels or Siding
1/2 & 5/8	(Permanently exposed to weather)
5/8 or 3/4	Concrete Forms (Multiple re-use)
5/8	Concrete Forms (Single use)
1/4 & 3/8	

PLYWOOD—WEIGHTS/100 SQUARE FEET

Thickness (In.)	Hardwood	Softwoods	Patterned Softwoods
1/4	825	750	—
5/16	—	—	950
3/8	1200	1125	—
7/16	—	—	1300
1/2	1600	1525	—
9/16	—	—	1675
11/16	—	—	2000
3/4	2325	2225	—
13/16	—	—	2375

POLYETHYLENE FILM—CLEAR BLACK (SIZES, WEIGHTS)

Thickness (In.)	Wt./1000 Sq. Ft.	Thickness (In.)	Maximum Width
.00075	3.6	.001 to .003	32'
.0001	4.8	.003 to .006	40'
.00125	6.0	.007	26'
.0015	7.2	.008	25'
.002	9.6	.009	22'- 6"
.0025	12.0	.010	20'-10"
.003	14.4	.011	20'
.004	19.2	.012	19'-2"
.005	24.0	.014	18'-4"
.006	28.8	.016	17'-6"
.007	33.6	.018	15'-10"
.008	38.4	.020	14'-2"
.009	43.2		
.010	48.0		
.012	57.6		
.014	67.2		
.015	72.0		
.016	76.8		
.018	86.4		
.020	96.0		

Pressure tape 2" × 100' rolls—Wt. 13 oz. per roll

POST BASES—STEEL

Size Post (In.)	A (In.)	Anchor Holes (In.)	Wt. Ea. (Lbs.)	Angle Size (In.)	Base Plate Thickness (In.)
4	3 3/4	9/16	14	2 × 2 × 1/4	1/4
6	5 3/4	9/16	19	2 × 2 × 1/4	1/4
8	7 3/4	9/16	25	2 × 2 × 1/4	1/4
10	9 3/4	9/16	31	2 × 2 × 1/4	1/4
12	11 3/4	9/16	36	2 × 2 × 1/4	1/4
14	13 3/4	9/16	63	2 × 2 1/2 × 1/4	5/16
16	15 3/4	13/16	74	2 × 2 1/2 × 1/4	5/16
18	17 3/4	13/16	83	2 × 2 1/2 × 1/4	5/16

Specific products vary

POST CAPS—STEEL

Post Size (In.)	Width Girder (In.)	Inside Width (In.)	Thickness (In.)	Bolts (In.)	Lags (In.)	Wt. Ea. (Lbs.)
6	6	5 3/4	1/4	1/2 × 7	3/8 × 1 1/2	31
8	8	7 3/4	1/4	1/2 × 9	3/8 × 3	39
10	10	9 3/4	1/4	5/8 × 11	1/2 × 3	48
12	12	11 3/4	1/4	3/4 × 13	1/2 × 3	57
14	14	13 3/4	5/16	3/4 × 15	1/2 × 4	66
16	16	15 3/4	5/16	1 × 17	1/2 × 5	76

LIGHT AND MEDIUM LOADS

Post Size (In.)	Width Girder (In.)	Inside Width (In.)	Thickness (In.)	Bolts (In.)	Lags (In.)	Wt. Ea. (Lbs.)
4	4	3 3/4	3/16	1/2 × 5	3/8 × 1 1/2	9
6	6	5 3/4	1/4	1/2 × 7	3/8 × 1 1/2	22
8	8	7 3/4	1/4	1/2 × 9	3/8 × 2	28
10	10	9 3/4	1/4	5/8 × 11	1/2 × 3	37
12	12	11 3/4	1/4	3/4 × 13	1/2 × 3	45

Also 3 and 4 way post caps

Various types and styles available

PROFIT—PERCENTAGE MARKUP REQUIRED

Gross Profit Desired	Add to Cost
50%	100%
40	66 2/3
35	53 4/5
30	42 6/7
25	33 1/3
20	25
15	17 2/3
12 1/2	14 2/7
10	11 1/9
8	8 2/3
5	5 1/4

R

R-VALUES—SOFTWOOD PLYWOOD

Panel Thickness (In.)	Thermal Resistance R
1/4	0.31
5/16	0.39
3/8	0.47
7/16	0.55
1/2	0.62
5/8	0.78
3/4	0.94
1	1.25
1 1/8	1.41

R-VALUES—SHEATHING ASSEMBLIES, INSULATED FRAME
Typical (Conductive Thermal Performance) * R-11 Mineral Fiber Batts with Vapor Barrier

Sheathing	1/2" Hard Board Siding	3/8" Plywood Siding	1/2" Aluminum Siding	1/2" Beveled, 8" Lapped Wood Siding	4" Brick Veneer Plus 3/4" – 1" Air Space Siding	1/2" Stucco Siding
1/2" Regular Density Wood Fiberboard	R-12.18 U-.0821	R-11.95 U-.0837	R-12.11 U-.0826	R-12.33 U-.0811	R-13.1 U-.0763	R-11.6 U-.0862
1/2" Intermediate Density Wood Fiberboard	R-12.06 U-.0826	R-11.84 U-.0884	R-12.0 U-.083	R-12.22 U-.0819	R-13.0 U-.0769	R-11.5 U-.087
1/2" Exterior Grade Plywood	R-11.4 U-.0877	R-11.18 U-.0894	R-11.34 U-.0882	R-11.56 U-.0865	R-12.36 U-.0809	R-10.82 U-.0924
1" Extruded Polystyrene Foam	R-16.1 U-.0622	R-15.9 U-.0630	R-16.0 U-.0624	R-16.2 U-.0616	R-17.0 U-.0589	R-15.5 U-.0644
1" Expanded Polystyrene Foam 1# PCF	R-14.9 U-.0672	R-14.7 U-.0682	R-14.8 U-.0675	R-15.0 U-.0666	R-15.8 U-.0633	R-14.3 U-.0698
1" Expanded Polystyrene Foam 1.5# PCF	R-15.21 U-.0657	R-15.0 U-.0667	R-15.15 U-.0660	R-15.36 U-.0651	R-16.12 U-.0620	R-14.66 U-.0682
3/4" Urethane-Isocyanurate Impermeable Facing	R-16.5 U-.0606	R-16.3 U-.0614	R-16.43 U-.0608	R-16.65 U-.0600	R-17.39 U-.0575	R-15.96 U-.0627
3/4" Urethane-Isocyanurate Permeable Facing	R-15.77 U-.0634	R-15.55 U-.0643	R-15.7 U-.0637	R-15.91 U-.0628	R-16.67 U-.0599	R-15.22 U-.0657

*At 75F mean temperature

R-VALUES SHEATHING ASSEMBLIES, INSULATED FRAME
Typical (Conductive Thermal Performance)* R-13 Mineral Fiber Batts with Vapor Barrier

Sheathing	1/2" Hard Board Siding	3/8" Plywood Siding	1/2" Aluminum Siding	1/2" Beveled, 8" Lapped Wood Siding	4" Brick Veneer Plus 3/4" – 1" Air Space Siding	1/2" Stucco Siding
1/2" Regular Density Wood Fiberboard	R-13.28 U-.0753	R-13.05 U-.0766	R-13.2 U-.0757	R-13.44 U-.0744	R-14.26 U-.0701	R-12.69 U-.0788
1/2" Intermediate Density Wood Fiberboard	R-13.17 U-.0759	R-12.94 U-.0773	R-13.1 U-.0763	R-13.33 U-.075	R-14.15 U-.0706	R-12.57 U-.0796
1/2" Exterior Grade Plywood	R-12.13 U-.0824	R-12.25 U-.0816	R-12.40 U-.0805	R-12.64 U-.0791	R-13.48 U-.0742	R-11.87 U-.0842
1" Extruded Polystyrene Foam	R-17.3 U-.0577	R-17.1 U-.0585	R-17.3 U-.0579	R-17.5 U-.0572	R-18.3 U-.0548	R-16.8 U-.0596
1" Expanded Polystyrene Foam 1# PCF	R-16.1 U-.0622	R-15.9 U-.0630	R-16.0 U-.0624	R-16.2 U-.0616	R-17.0 U-.0587	R-15.5 U-.0644
1" Expanded Polystyrene Foam 1.5# PCF	R-16.43 U-.0608	R-16.31 U-.0613	R-16.39 U-.0610	R-16.58 U-.0603	R-17.0 U-.0587	R-15.86 U-.0630
3/4" Urethane Isocyanurate Impermeable Facing	R-17.75 U-.0563	R-17.62 U-.0567	R-17.69 U-.0565	R-17.81 U-.0560	R-18.67 U-.0536	R-17.18 U-.0582
3/4" Urethane-Isocyanurate Permeable Facing	R-17.0 U-.0588	R-16.79 U-.0595	R-16.94 U-.059	R-17.15 U-.0583	R-17.93 U-.0558	R-16.56 U-.0603

* At 75F Mean Temperature.

R-VALUES—SUGGESTED MINIMUMS

Feature	Zone A	Zone B	Zone C	Zone D	Zone E
FOR HOMES HEATED WITH ELECTRIC RESISTANCE HEAT					
1. Ceiling Insulation	R-19	R-22	R-30	R-30	R-38
2. Wall Insulation	R-11	R-13	R-19	R-19	R-19
3. Floors Over Unheated Spaces	none	R-11	R-19	R-19	R-19
4. Foundation Walls of Heated Spaces	none	none	R-6	R-11	R-11
5. Slab Foundation Perimeter	none	R-5	R-7.5	R-7.5	R-7.5
FOR HOMES HEATED WITH OIL, GAS OR HEAT PUMPS					
1. Ceiling Insulation	R-19	R-19	R-26	R-30	R-38
2. Wall Insulation	R-11	R-11	R-13	R-13	R-19
3. Floors Over Unheated Spaces	none	none	R-11	R-11	R-19
4. Foundation Walls of Heated Spaces	none	none	R-6	R-11	R-11
5. Slab Foundation Perimeter	none	R-2	R-5	R-5	R-7.5

RAFTERS—LENGTHS

Width of Building	FOURTH PITCH					THIRD PITCH					HALF PITCH				
	Length of Rafter		From Plate to Comb		Area of Two Gables	Length of Rafter		From Plate to Comb		Area of Two Gables	Length of Rafter		From Plate to Comb		Area of Two Gables
	(Ft.)	(In.)	(Ft.)	(In.)		(Ft.)	(In.)	(Ft.)	(In.)		(Ft.)	(In.)	(Ft.)	(In.)	
6	3	4	1	6	9	3	7	2	0	12	4	3	3	0	18
7	3	11	1	9	12	4	0	2	4	16	5	0	3	6	25
8	4	6	2	0	16	4	10	2	8	21	5	8	4	0	32
9	5	0	2	3	20	5	5	3	0	27	6	5	4	6	41
10	5	7	2	6	25	6	0	3	4	33	7	1	5	0	50
12	6	8	3	0	36	7	2	4	0	48	8	6	6	0	72
14	7	10	3	6	49	8	5	4	8	65	9	11	7	0	98
16	9	0	4	0	64	9	7	5	4	85	11	4	8	0	128
18	10	1	4	6	81	10	10	6	0	108	12	9	9	0	162
20	11	2	5	0	100	12	0	6	8	133	14	2	10	0	200
22	12	4	5	6	121	13	2	7	4	161	15	7	11	0	242
24	13	5	6	0	144	14	5	8	0	192	17	0	12	0	288
26	14	6	6	6	169	15	7	8	8	225	18	6	13	0	338
28	15	8	7	0	196	16	10	9	4	261	19	11	14	0	392
30	16	9	7	6	225	18	0	10	0	300	21	4	15	0	450
32	17	11	8	0	256	19	2	10	8	341	22	9	16	0	512

RAFTERS—MAXIMUM SPAN LIMITS
Live Load 30 Pounds/Square Foot

Size of Rafters (In.)	Spacing of Rafters (In.)	White Pine	Hemlock	Douglas Fir	Long Leaf Yellow Pine
2 × 4	16	7′ – 9″	8′ – 7″	9′ – 0″	9′ – 7″
2 × 4	24	6′ – 5″	7′ – 0″	7′ – 5″	8′ – 0″
2 × 6	16	11′ – 10″	13′ – 2″	13′ – 9″	14′ – 9″
2 × 6	24	9′ – 9″	10′ – 10″	11′ – 5″	12′ – 3″
2 × 8	16	15′ – 8″	17′ – 5″	18′ – 2″	19′ – 6″
2 × 8	24	13′ – 0″	14′ – 5″	15′ – 1″	16′ – 3″

Check local codes

REGISTERS—AIR (SIZES)

Pipe Size (In.)	Free Area (In.)	Register Size, Floor (In.)	Register Size, Baseboard (In.)
8	50	8 × 10	8 × 10 – 2 1/4
9	64	9 × 12	8 × 12 – 2 1/4
10	79	10 × 12	9 × 12 – 3 1/4
12	113	12 × 14	11 × 13 – 4 1/4
14	154	14 × 16	

REINFORCING BAR STEEL—ROUND & SQUARE
(WEIGHTS & AREAS)

Bar No.	Size of Bar (In.)	Type	Area (Sq. In.)	Pounds/ (Ft.)	Lin. Ft./ (Ton)	Diam. (In.)	Perim. (In.)
2	1/4	Round	.049	.167	12,000	.250	.786
3	3/8	Round	.111	.376	5,320	.375	1.178
4	1/2	Round	.196	.668	3,000	.500	1.571
5	5/8	Round	.307	1.043	1,912	.625	1.963
6	3/4	Round	.442	1.502	1,331	.750	2.356
7	7/8	Round	.601	2.044	978	.875	2.749
8	1	Round	.785	2.670	749	1.000	3.142
9	1	Square	1.000	3.400	588	1.128	3.544
10	1 1/8	Square	1.266	4.303	465	1.270	3.990
11	1 1/4	Square	1.563	5.313	376	1.410	4.430

ROMAN NUMERALS

I	....	1
II	....	2
III	....	3
IV	....	4
V	....	5
VI	....	6
VII	....	7
VIII	....	8
IX	....	9
X	....	10
XI	....	11
XII	....	12
XIII	....	13
XIV	....	14
XV	....	15
XVI	....	16
XVII	....	17
XVIII	....	18
XIX	....	19
XX	....	20
XXX	....	30
XL	....	40
L	....	50
LX	....	60
LXX	....	70
LXXX	....	80
XC	....	90
C	....	100
CC	....	200
CCC	....	300
CCCC	....	400
D	....	500
DC	....	600
DCC	....	700
DCCC	....	800
CM	....	900
M	....	1000
MCM	....	1900
MM	....	2000

ROOF—AREA DRAINED BY LEADER PIPE

Size of Pipe (In.)	Area of Pipe (Sq. Ft.)	Roof Area Drained (Sq. In.)
3	7.06	1,050
4	12.56	1,875
5	19.63	2,950
6	28.27	4,200
8	50.26	7,500

Based on average rainfall conditions

ROOF AREAS—PITCH

1/4	Pitch	Add to square area	12%
1/3	Pitch	Add to square area	20%
1/2	Pitch	Add to square area	42%
3/8	Pitch	Add to square area	25%
5/8	Pitch	Add to square area	60%
3/4	Pitch	Add to square area	80%

ROOF CEMENTS

Roof Surface	Coverage (Gals./Sq.)
Concrete	2
Gravel	4
Metal (Corr.)	2
Metal (Sm.)	1 1/2
Smooth Asp.	2 1/2

40 to 50 lbs. of plastic will cover 1 sq. 1/16-inch thick.

ROOF CEMENTS

Sizes (Gal.)	Weights	
	Liquid	Plastic
1	8	10
5	40	50
30	250	300
55	450	550

ROOFING ASPHALT

- Packed: 100-pound rolls, bags
- Weight: 8.8 pounds/gallon when melted
- Melting Point: 165 – 175°F

ROOFING—BUILT-UP

Specifications for Fireproof Decks
Material Weights: Pounds/Square Feet

Years Bond	Underwriters Classification	Roof Incline (In./Ft.)	Top Surface	Plies of Felt	Layers of Bitumen	Bitumen Primer	Sheathing Paper	Base Sheet	Asphalt Felt	Tarred Felt	Rex Construction	Asphalt	Pitch	Viskalt	Gravel	Slag	Smooth	Mineral Surfaced	Gravel Surfaced	Slag Surfaced
20	A	1/4" up to 3"	Gravel or slag	Four #15 felts (asphalt)	5	9			60			190			400	300			659	559
20	A	1/8" up to 2"	Gravel or slag	Four #15 felts (tar)	5					60			200		400	300			660	560
15	A	1/4" up to 3"	Gravel or slag	Three #15 felts (asphalt)	4	9			45			160			400	300			614	514
15	A	1/8" up to 2"	Gravel or slag	Three #15 felts (tar)	4	9				45			175		400	300			620	520
15	C	1" up to 6"	Mineral	Two #15 felts (asphalt)	4	9			30		110	120						269		
10	C	1" up to 6"	Mineral	One #15 felts (asphalt)	3	9			15		110	90						224		
10	C	1/2" up to 1"	Mineral	Three #15 felts (asphalt)	4	9			45					140			194			
10		2" up to 4"	Slag	Three #15 felts	4	9			45			130				275				459
20	A	1/4" up to 3"	Gravel or Slag	One #30 felt (asphalt) Three #15 felts (asphalt)	4			30	45			150			400	300			625	525
20	A	1/8" up to 2"	Gravel or Slag	Five #15 felts (tar) One sheathing paper	4		5			75			150		400	300			630	530
15	A	1/4" up to 3"	Gravel or Slag	One #30 felt (asphalt) Two #15 Felts (asphalt)	3			30	30			120			400	300			580	480
15	A	1/8" up to 2"	Gravel or Slag	Four #15 felts (tar) One sheathing paper	3		5			60			125		400	300			590	490
15	C	1" up to 6"	Mineral	Two #15 felts (asphalt)	3				30		110	90						235		
10	C	1" up to 6"	Mineral	Two #15 felts (asphalt)	2				30		110	60						200		
10	C	1/2" up to 1"	Mineral	One #30 felt (asphalt) Two #15 felts (asphalt)	3			30	30					100			160			
10	C	2" up to 4"	Slag	Four #15 felts	4				60			125				275				460

Specifications may vary in different localities

ROOFING—BUILT-UP (NAILING REQUIREMENTS)

Roof Incline	Nail Type	Nail Spacing
Up to $\frac{1}{2}$" per ft.	$\frac{7}{8}$" or 1"	10' apart each ply
Up to 1" per ft.	$\frac{7}{8}$" or 1"	8' apart each ply
Up to 1 $\frac{1}{2}$" per ft.	$\frac{7}{8}$" or 1"	5' apart each ply
Up to 2" per ft.	$\frac{7}{8}$" or 1"	3' apart each ply
Up to 4" per ft.	$\frac{7}{8}$" or 1"	2' apart each ply
Over 4" per ft.	$\frac{7}{8}$" or 1"	1' apart each ply

Check local requirements

ROOFING—BUILT-UP (UNDER TILE FLOORS, DECKS)

Incline	Felts	Moppings	Weights (Lbs.)			Total Wt. Sq. (Lbs.)
			Primer	Felt	Asphalt	
Not over 1" per ft.	5 – 15	6	10	75	210	295

ROOFING—CLAY TILE

Styles	Exposure	Wt./ Sq. (Lbs.)
Rectangular) Interlocking)	7 $\frac{7}{8}$ × 10 $\frac{3}{4}$	800#
Rectangular) Interlocking)	8 × 8	850
Shingle	7 $\frac{1}{2}$ × 6 $\frac{1}{2}$	1500
Shingle	7 × 5	1750
Spanish	8 × 10 $\frac{1}{4}$	900
Roman	12 × 10	1300
Greek	12 × 10	1350
	19 × 15	
Tapered Mission	11 $\frac{1}{2}$ × 11 $\frac{1}{4}$	1325
Barrel Mission	11 $\frac{1}{2}$ × 11 $\frac{1}{2}$	1250
	× 15	
French	8 $\frac{1}{8}$ × 13 $\frac{1}{4}$	950

ROOFING—CONCRETE TILE

Type	Shingle Size (In.)	Approx. Exp. (In.)	No./ Sq.	Wt./ Sq.
Spanish	9 3/4 × 15 1/8	8 × 12	130	800
French	" "	" "	150	950
Slate &	" "	" "		
French	" "	" "	144	1050

ROOFING—COPPER (STANDING-SEAM)

Sheet Sizes (In.)	Width between Seams (In.)	Gauge (Oz.)
16 × 72	13 3/4	10
24 × 96	21 1/4	16
28 × 96	25 1/4	20
32 × 96	29 1/4	24

ROOFING—CORRUGATED STEEL (NUMBER OF SHEETS/SQUARE)
No Allowance for Laps

Length of Sheet (In.)	2 1/2 In. Corrugations (Sheet 26 In. Wide)	1 1/4 In. Corrugations (Sheet 25 In. Wide)
60	9.231	9.600
72	7.692	8.000
84	6.593	6.857
96	5.769	6.000
108	5.128	5.333
120	4.616	4.800
144	3.846	4.000

ROOFING—CORRUGATED STEEL
(SQUARE FEET/SHEET)

Length of Sheet (In.)	2 1/2 In. Corrugations (Sheet 26 In. Wide)	1 1/4 In. Corrugations (Sheet 25 In. Wide)
60	10.833	10.416
72	13.000	12.500
84	15.166	14.583
96	17.333	16.666
108	19.500	18.750
120	21.666	20.833
144	26.000	25.000

ROOFING—FIBERGLASS ROLL (MINERAL-SURFACED)

Types	Wt./ Sq.	Width (In.)	Length (Ft.)	Sq. Ft./ Roll
M. S. Roll	105	36	36	108
Selvedged	110	36	36	54

Both types require approx. 175 1-inch nails per sq.

Available in various colors and color combinations in many areas. Other types may also be available.

ROOFING—FIBERGLASS SHINGLES (CLASS A)

Types	Size (In.)	Roof Pitch Min. (In.)	Shgls./ Sq.	Bdls./ Sq.	Exp. (In.)	Wt. (Lbs.) Sq.
Random Tab Strip	13 5/8 × 39 5/8	4	72	3	5	275
Staggered Edge Strip	12 1/3 × 36	2	89	3	4 1/4	275
" " "	13 7/16 × 39 3/8	2	72	3	5 1/8	265
Slab Strip	12 × 36	4	80	3	5	275
3 Tab Strip	" "	2	80	3	5	230
Dbl. Coverage Lock	21 1/2 × 19	4	108	3		230
French Reroofing	16 × 16	4	82	2		160

Require approx. 250 1-inch nails per sq. or approved staples.
Reroofing type requires approx. 175 1 3/4-inch nails or heavy long staples.

ROOFING—RED CEDAR SHAKES (COVERAGE)

Shake Type, Length and Thickness	Approximate Coverage (in Sq. Ft.) of One Square, When Shakes Are Applied with 1/2" Spacing, at Following Weather Exposures, in Inches (h):					
	5	7 1/2	8 1/2	10	11 1/2	16
18" × 1/2" Handsplit-and-Resawn Mediums (a)	55(b)	75(c)	85(d)	100	....	....
18" × 3/4" Handsplit-and-Resawn Heavies (a)	55(b)	75(c)	85(d)	100	....	....
18" × 5/8" Tapersawn	55(b)	75(c)	85(d)	....	....	....
24" × 3/8" Handsplit	....	75(e)	85	100(f)	115(d)	....
24" × 1/2" Handsplit-and-Resawn Mediums	....	75(b)	85	100(c)	115(d)	....
24" × 3/4" Handsplit-and-Resawn Heavies	....	75(b)	85	100(c)	115(d)	....
24" × 5/8" Tapersplit	....	75(b)	85	100(c)	115(d)	....
24" × 1/2" Tapersplit	....	75(b)	85	100(c)	115(d)	....
18" × 3/8" True-Edge Straight-Split	....	....	....	....	....	112(g)
18" × 3/8" Straight-Split	65(b)	90	100(d)	....	....	....
24" × 3/8" Straight-Split	....	75(b)	85	100	115(d)	....
15" Starter-Finish Course	Use supplementary with shakes applied not over 10" weather exposure.					

(a) 5 bundles will cover 100 sq. ft. roof area when used as a starter-finish course at 10" weather exposure; 6 bundles will cover 100 sq. ft. wall area at 8 1/2" exposure; 7 bundles will cover 100 sq. ft. roof area at 7 1/2" weather exposure; see footnote (h).

(b) Maximum recommended weather exposure for 3-ply roof construction.

(c) Maximum recommended weather exposure for 2-ply roof construction.

(d) Maximum recommended weather exposure for single-coursed wall construction.

(e) Maximum recommended weather exposure for application on roof pitches between 4-in-12 and 8-in-12.

(f) Maximum recommended weather exposure for application on roof pitches of 8-in-12 and steeper.

(g) Maximum recommended weather exposure for double-coursed wall construction.

(h) All coverage based on 1/2" spacing between shakes.

255

Red Cedar—Shakes, see page 257.

ROOFING—RED CEDAR SHINGLES (COVERAGE)

Approximate Coverage of One Square (4 Bundles) of Shingles Based on Following Weather Exposures

Length and Thickness	3½"	4"	4½"	5"	5½"	6"	6½"	7"	7½"	8"	8½"	9"	9½"
16" × 5/2"	70	80	90	100*	110	120	130	140	150‡	160	170	180	190
18" × 5/2 1/4"	....	72½	81½	90½	100*	109	118	127	136	145½	154½‡	163½	172½
24" × 4/2"	....	....	....	....	....	80	86½	93	100*	106½	113	120	126½

10"	10½"	11"	11½"	12"	12½"	13"	13½"	14"	14½"	15"	15½"	16"
200	210	220	230	240†	....	....	....	....	....	....	....	....
181½	191	200	209	218	227	236	245½	254½	....	....	....	....
133	140	146½	153‡	160	166½	173	180	186½	193	200	206½	213†

*Maximum exposure recommended for roofs.

†Maximum exposure recommended for double-coursing No. 1 grades on sidewalls.

‡Maximum exposure recommended for single-coursing No. 1 and No. 2 grades on sidewalls.

ROOFING—RED CEDAR SHINGLES (MAXIMUM RECOMMENDED EXPOSURE)

Pitch	No. 1 Blue Label			No. 2 Red Label			No. 3 Black Label		
	16"	18"	24"	16"	18"	24"	16"	18"	24"
3 in 12 to 4 in 12	3¾"	4¼"	5¾"	3½"	4"	5½"	3"	3½"	5"
4 in 12 and steeper	5"	5½"	7½"	4"	4½"	6½"	3½"	4"	5½"

ROOFING—RED CEDAR SHAKES (STANDARD GRADES)

Grade	Length and Thickness	18″ Pack**	
		Courses/ Bdl.	Bdls./ Sq.
No. 1 Handsplit & Resawn	15″ Starter-Finish	$9/9$	5
	18″ × $1/2$″ Mediums	$9/9$	5
	18″ × $3/4$″ Heavies	$9/9$	5
	24 × $3/8$″	$9/9$	5
	24″ × $1/2$″ Mediums	$9/9$	5
	24″ × $3/4$″ Heavies	$9/9$	5
No. 1 Tapersawn	24″ × $5/8$″	$9/9$	5
	18″ × $5/8$″	$9/9$	5
No. 1 Tapersplit	24″ × $1/2$″	$9/9$	5
		20″ Pack	
No. 1 Straight- split	18″ × $3/8$″ True-Edge*	14 Straight	4
	18″ × $3/8$″	19 Straight	5
	24″ × $3/8$″	16 Straight	5

* Exclusively sidewall product, with parallel edges.
**Pack used for majority of shakes.

ROOFING—RED CEDAR SHINGLES (STANDARD GRADES)

Grade	Length	Thickness (at Butt)	No. of Courses/ Bundle	Bdls./Cartons /Square
No. 1 Blue Label	16″ (Fivex)	.40″	$20/20$	4 bdls.
	18″ (Perfections)	.45″	$18/18$	4 bdls.
	24″ (Royals)	.50″	$13/14$	4 bdls.
No. 2 Red Label	16″ (Fivex)	.40″	$20/20$	4 bdls.
	18″ (Perfections)	.45″	$18/18$	4 bdls.
	24″ (Royals)	.50″	$13/14$	4 bdls.
No. 3 Black Label	16″ (Fivex)	.40″	$20/20$	4 bdls.
	18″ (Perfections)	.45″	$18/18$	4 bdls.
	24″ (Royals)	.50″	$13/14$	4 bdls.
No. 4 Under- coursing	16″ (Fivex)	.40″	$14/14$ or $20/20$	2 bdls. 4 bdls.
	18″ (Perfections)	.45″	$14/14$ or $18/18$	2 bdls. 4 bdls.
No. 1 or No. 2 Rebutted- Rejointed	16″ (Fivex)	.40″	$33/33$	1 carton
	18″ (Perfections)	.45″	$28/28$	1 carton
	24″ (Royals)	.50″	$13/14$	4 bdls.

ROOFING—SLATE

Sizes of Slate (In.)	Number/ Square	Exposed When Laid 3 In. Lap	Nails to Square 3d (Lbs.)
26 × 14	89	11 1/2″	1
24 × 16	86	10 1/2	1
24 × 14	98	10 1/2	1
24 × 13	106	10 1/2	1
24 × 11	125	10 1/2	1 1/2
24 × 12	114	10 1/2	1
22 × 14	108	9 1/2	1
22 × 13	117	9 1/2	1
22 × 12	126	9 1/2	1
22 × 11	138	9 1/2	2
22 × 10	152	9 1/2	2
20 × 14	121	8 1/2	1
20 × 13	132	8 1/2	2
20 × 12	141	8 1/2	2
20 × 11	154	8 1/2	2
20 × 10	170	8 1/2	2
20 × 9	189	8 1/2	2
18 × 14	137	7 1/2	2
18 × 13	148	7 1/2	2
18 × 12	160	7 1/2	2
18 × 11	175	7 1/2	2
18 × 10	192	7 1/2	2
18 × 9	213	7 1/2	2
16 × 14	160	6 1/2	2
16 × 12	184	6 1/2	2
16 × 11	201	6 1/2	2
16 × 10	222	6 1/2	3
16 × 9	246	6 1/2	3
16 × 8	277	6 1/2	3
14 × 12	218	5 1/2	2 1/2
14 × 11	238	5 1/2	3
14 × 10	261	5 1/2	3
14 × 9	291	5 1/2	3
14 × 8	327	5 1/2	4
14 × 7	374	5 1/2	4
12 × 10	320	4 1/2	4
12 × 9	355	4 1/2	4
12 × 8	400	4 1/2	5
12 × 7	457	4 1/2	5
12 × 6	533	4 1/2	6
11 × 8	450	4	5
11 × 7	515	4	6
10 × 8	515	3 1/2	6
10 × 7	588	3 1/2	7
10 × 6	686	3 1/2	8

Weights/Sq.

3/16″	750	1/2″	1880
1/4″	1050	3/4″	2500
3/8″	1300	1″	3500

Roofing—Types & Characteristics, see page 261.

ROOFING & SIDING—CORRUGATED ALUMINUM SHEETS

Width (In.)	Sheet Thickness	Total Thickness (In.)	Sheet Length (Feet)	Wt./ Sq. Ft. (Pounds)	Sq. Ft./ Sheet	Pounds/ Sheet	No. of Sheets/ 100 Sq. Ft.	Corrugations Crown to Crown (In.)	Coverage (In.)
35	.032	7/8	5	.56	14.58	8.17	6.85	2.67	32
"	"	"	6	"	17.50	9.80	5.71	"	"
"	"	"	7	"	20.42	11.43	4.89	"	"
"	"	"	8	"	23.33	13.07	4.29	"	"
"	"	"	9	"	26.25	14.70	3.80	"	"
"	"	"	10	"	29.17	16.33	3.42	"	"
"	"	"	11	"	32.08	17.95	3.11	"	"
"	"	"	12	"	35.00	19.60	2.85	"	"

Also made in .024-inch, 42 lbs./sq. ft.

Other types available

ROOFING & SIDING—GALVANIZED STEEL SHEETS
Weights/Square

Gauge	28	26	24	Painted			
				22	20	18	16
2", 2 1/2", 3" Corrugated, 26" wide	68	82	109	136	163	216	270
2 1/2", Corrugated, 27 1/2" wide	69	83	110	137	165	219	274
1 1/4", Corrugated, 26" wide	68	82	109	136	163	...	...
5/8", Corrugated, 25" wide	71	85	...	...	...	...	...
2-V Crimp Roofing, 24" wide	69	82	109	137	164	...	...
3-V Crimp Roofing, 24" wide	70	84	112	139	167	...	...
5-V Crimp Roofing, 24" wide	74	89	118	...	...	...	...
Pressed Standing Seam Roofing, 24" wide	70	84	112	...	...	...	...
Roll Roofing, 26 1/2" wide, no cleats	72	86	...	...	...	...	...
Roll and Cap Roofing, 26" wide	76	91	120	...	...	...	...
Plain, Beaded Ceiling, 24" wide	69	82	109	...	...	...	...
Weather Board Siding, 24" wide	74	89	118	...	...	...	...
Plain Pressed Brick Siding, 28" x 60"	64	76	...	...	...	...	...
Rock Face Brick or Stone Siding, 28" x 60"	64	76	...	...	...	...	...

For plain sheets, deduct one pound per sq.
For galvanized products, add approx. 10%

Numerous styles available

ROOFING—TYPES & CHARACTERISTICS

Type	Fire Rating	Minimum Slope	Weight (Lbs.)/ Square
Aluminum Shingles	A	3/12	35 – 65
Asphalt Roll Roofing	C to A	1/12	45 – 110
Asphalt Sheets, Corrugated	C	2/12	81 – 122
Asphalt Shingles, Felt	C to A	2/12	235 – 390
Asphalt Shingles, Fiberglass	A	2/12	210 – 390
Cedar Shakes	C*	4/12	200 – 300
Cedar Shingles	C*	4/12	160 – 250
Copper (16 oz.)	A	1/12	100
Mineral Fiber Shingles	A	3/12	325 – 500
Slate Shingles	A	4/12	700 – 3500
Steel, Corrugated	A	1/12	120 – 170
Tile, Clay	A	3/12	750 – 1500
Tile, Concrete	A	4/12	850 – 1050
Wood-fiber Sheets	C*	3.5/12	260 – 280

*When treated with fire-retardant chemicals; otherwise, Unclassified

ROPE—WIRE (STANDARD CAST STEEL)
6 Strands, 19 Wires/Strand Hemp Core

Rope Diam. (In.)	Approx. Weight/ Foot (Lbs.)	Approx. Circum- ference (In.)	Breaking Strength (Tons)
1/4	0.10	3/4	2.1
5/16	.16	1	3.2
3/8	.23	1 1/8	4.5
7/16	.31	1 3/8	6.0
1/2	.40	1 5/8	7.7
9/16	.51	1 3/4	9.6
5/8	.63	2	11.8
3/4	.90	2 3/8	16.8
7/8	1.23	2 3/4	22.8
1	1.60	3 1/8	29.5
1 1/8	2.03	3 1/2	37.0
1 1/4	2.50	3 7/8	46.0

ROPE—MANILA

Diameter (In.)	Weight (Lbs.) of 1200 Ft.	Strength (Lbs.)
1/4	22	620
5/16	29	1,000
3/8	44	1,275
7/16	65	1,875
1/2	90	2,400
9/16	125	3,300
5/8	160	4,000
3/4	198	4,700
13/16	234	5,600
7/8	270	6,500
1	324	7,500
1 1/16	378	8,900
1 1/8	432	10,500
1 1/4	504	12,500
1 5/16	576	14,000
1 3/8	648	15,400
1 1/2	720	17,000
1 9/16	810	18,400
1 5/8	900	20,000
1 3/4	1,080	25,000
2	1,296	30,000
2 1/8	1,512	33,000
2 1/4	1,764	37,000
2 1/2	2,016	43,000
2 5/8	2,304	50,000
2 7/8	2,590	56,000
3	2,915	62,000
3 1/8	3,240	68,000
3 1/4	3,600	75,000

S

SAND

- A cu. yd. of sand weighs 2,700 lbs.
- A ton of sand is approximately $^3/_4$ cu. yd.
- A cu. ft. of sand weighs approximately 100 lbs.
- A 12-qt. bucket holds approximately 40 lbs. of sand.
- A ton of sand contains approximately 20 cu. ft.
- Average shovel of sand weighs 15 lbs.

SAND—CONCRETE

Sieve Size	% Passing
$^3/_8$ ″	100
$^1/_4$ ″	95 – 100
$^1/_8$ ″	70 – 95
$^1/_{16}$″	45 – 80
30 Mesh	25 – 60
50 Mesh	10 – 30
100 Mesh	1 – 10

SAND—MASONRY

Sieve Size	Percentages Passing
$^1/_4$ ″	100
$^1/_8$ ″	95 to 100
$^1/_{16}$″	60 to 100
30 Mesh	30 to 70
50 Mesh	15 to 35
100 Mesh	0 to 15

SAND—PLASTER

Sieve Size	% Retained On	
	Max.	Min.
$^1/_4$″	0	0
$^1/_8$″	10	0
30 Mesh	80	15
50 Mesh	95	70
100 Mesh	100	95

SCREWS—GALVANIZED LAG

Size (In.)	Weight/ 100 (Lbs.)
$3/8 \times 2 \ 1/4$	9
$3/8 \times 3$	10
$3/8 \times 3 \ 1/2$	11
$3/8 \times 4$	12
$3/8 \times 4 \ 1/2$	13
$3/8 \times 5$	14
$1/2 \times 2 \ 1/2$	16
$1/2 \times 3$	19
$1/2 \times 3 \ 1/2$	21
$1/2 \times 4$	23
$1/2 \times 4 \ 1/2$	26
$1/2 \times 5$	28
$1/2 \times 6$	32
$1/2 \times 7$	37

SCREWS—(SIZES)

Ga.	Approx. Dia. (In.)
0	$1/16$
1	$5/64$
2	$5/64$
3	$3/32$
4	$7/64$
5	$1/8$
6	$9/64$
7	$5/32$
8	$5/32$
9	$11/64$
10	$3/16$
11	$13/64$
12	$7/32$
14	$15/64$
16	$17/64$
18	$19/64$
20	$21/64$
24	$3/8$

SCREWS—LAG

Lag Screw Size, Dia. (In.)	Shield Length (In.)	Drill Size (In.)
$1/4$	1 to 1 $1/2$	$1/2$
$5/16$	1 $1/4$ to 1 $3/4$	$1/2$
$3/8$	1 $3/4$ to 2 $1/2$	$5/8$
$1/2$	2 to 3	$3/4$
$5/8$	2 to 3 $1/2$	$7/8$
$3/4$	2 to 3 $1/2$	1

Sewer Joints—Poured Asphalt, see page 266.
Sheathing—Gypsum, see page 267.

SHUTTERS—WOOD

Size of Pair	2 Lt. Glass Size
2 – 0 1/2 × 3 – 4	20 × 16
4 – 0	20
4 – 8	24
5 – 0	26
2 – 4 1/2 × 3 – 4	24 × 16
3 – 8	18
4 – 0	20
4 – 8	24
5 – 0	26
2 – 6 1/2 × 4 – 8	26 × 24
5 – 0	26
2 – 8 1/2 × 3 – 4	28 × 16
3 – 8	18
4 – 0	20
4 – 4	22
4 – 8	24
5 – 0	26
2 – 10 1/2 × 3 – 4	30 × 16
4 – 0	20
4 – 4	22
4 – 8	24
5 – 0	26
5 – 4	28
5 – 8	30
3 – 0 1/2 × 3 – 4	32 × 16
4 – 0	20
4 – 8	24
5 – 0	26
3 – 4 1/2 × 3 – 4	36 × 16
4 – 0	20
4 – 8	24
5 – 0	26
3 – 8 1/2 × 4 – 8	40 × 24

Average Wt. 16 lbs./ pr.

SEWER JOINTS—POURED ASPHALT

Pipe Size (In.)	Lbs. Asphalt/ 100 Jts. Bell and Spigot Jts.	T & G Joints
4	50	
6	115	
8	180	
10	200	
12	300	160
15	400	270
18	500	400
21	700	600
24	850	800
27	1200	1200
30	1300	1400
33	1600	1425
36	1900	1900
42		2500
48		3000
54		4502
60		6000
72		9100

SIDING—ALUMINUM

Type	Width	Length
Horiz. V-Groove	8″	12′–16′
Horiz. Plain	12″	8′
Bd. & Batten	12″	12′–16′
Twin 4-Inch	8″	12′–16′
Vertical	10″	9′ 3″ & 10′
Horiz. Panel	4′	8′
Vert. Panel	4′	9′
Wthrbd. Plain	24″ panels	8′, 10′ 12′
Wthrbd. Simulated	25 1/2″	8′, 10′, 12′

Sizes, types may vary. Double 4- and 5-inch available in some areas. Vent and weep holes in most types.

SHEATHING—GYPSUM (TYPICAL PRODUCT SIZES)

Type	Thickness (In.)	Length (Ft.)	Width (In.)	Edge
Regular	3/8	8	24	T&G
	3/8	8, 9	48	Square
	4/10	8, 9	24	T&G
	4/10	8, 9	48	Square
	1/2	8, 9	24	T&G
	1/2	8, 9	48	Square
	5/8	8, 9	24	T&G
	5/8	8, 9	48	Square
Type X	5/8	8	24	T&G
	5/8	8, 9	48	Square

SIDING—BEVEL
(METAL CORNERS)

Siding Size (In.)	Size (In.)	Weight (Lbs.)/ 100
1/2 × 6	1/2 × 5 1/2	10
1/2 × 8	1/2 × 7 1/2	15
3/4 × 8	3/4 × 7 1/2	16
3/4 × 10	3/4 × 9 1/2	20

SIDING—LAP OR BEVEL

Lap Sizes	Add for Lap	Wt. (Lbs.)/ 1000 Ft.
1/2 × 6	26%	600
1/2 × 8	18%	600
3/4 × 8	18%	900
3/4 × 10	14%	900
V JOINT		
3/4 × 7 1/4	12%	2000

SIDING—HARDBOARD (APPLICATION)

Siding	Maximum Framing Spacing	Nail Size(1)	Nail Spacing		Joint Gap	Gap Around Openings
			Shear Strength Not Provided (2)	Shear Strength Req'd(3)		
LAP SIDING DIRECT TO STUDS	16"o.c.	8d	16"o.c.	Not Applicable	Moderate Contact	3/16"
OVER SHEATHING	16"o.c.	10d	16"o.c.	Not Applicable	Moderate Contact	3/16"
SQUARE EDGE PANEL SIDING DIRECT TO STUDS	24"o.c.	6d	6"o.c. Edges 12"o.c. Intermed.	4"o.c. Edges 8"o.c. Intermed.	1/8"	3/16"
OVER SHEATHING	24"o.c.	8d	6"o.c. Edges 12"o.c. Intermed.	N/A	1/8"	3/16"
SHIPLAP EDGE PANEL SIDING DIRECT TO STUDS	16"o.c. (4)	6d	6"o.c. Edges 12"o.c. Intermed.	4"o.c. Edges 8"o.c. Intermed.	Moderate Contact	3/16"
OVER SHEATHING	16"o.c. (4)	8d	6"o.c. Edges 12"o.c. Intermed.	N/A	Moderate Contact	3/16"

(1) Nail length must accommodate sheathing and penetrate framing 1 1/2 inches.
(2) Shear resistance provided by sheathing or corner braces.
(3) Shear resistance provided by siding.
(4) Some manufacturers permit application to studs spaced 24"o.c. and should be consulted for their recommendations.

SIDING—LAP OR BEVEL

Measured Width (In.)	Actual Width (In.)	Exposed to Weather	Add for Lap & Waste	Bd. Ft./ 100 Sq. Ft. of Surface	Nails	
					Size	No./1000 16" Ctrs
4	3 1/4	2 3/4	46%	151	6d	22
6	5 1/4	4 3/4	26%	131	6d	15
8	7 1/4	6 3/4	18%	123	6d	12
10	9 1/4	8 3/4	14%	119	7d	50
12	11 1/4	10 3/4	12%	117	8d	18

SIDING—RED CEDAR SHAKES (STANDARD GRADE)

Grade	Length	Thickness (at Butt)	No. Courses/ Carton	Cartons/ Square*
NO. 1 BLUE LABEL	16"	.40"	16/17	2 ctns.
	18"	.45"	14/14	2 ctns.
	24"	.50"	12/12	2 ctns.

*Also marketed in one-carton squares.

SIDING—SHAKES & SHINGLES RED CEDAR (WEATHER EXPOSURE)

Shingle Length	Maximum Weather Exposure		Shake Length and Type	Maximum Weather Exposure	
	Single-Course	Double-Course		Single-Course	Double-Course
16"*	7 1/2"	12"	18" resawn	8 1/2"	14"
18"*	8 1/2"	14"	24" resawn	11 1/2"	18"
24"*	11 1/2"	16"	24" tapersplit	11 1/2"	18"
			18" straightsplit	8 1/2"	16"
			18" tapersawn	8 1/2"	14"
*Includes machine-grooved shakes.			24" tapersawn	11 1/2"	18"

Siding—Redwood Bevel, see page 272.

SIDING—REDWOOD TONGUE AND GROOVE

Size (Nom.)	A	B	C	Pattern No.
SQUARE EDGES				
3/4×3	2 1/2	2 1/4	11/16	631
3/4×4	3 1/2	3 1/4	11/16	632
3/4×6	5 1/2	5 1/4	11/16	633
1×3	2 1/2	2 1/4	3/4	131
1×4	3 1/2	3 1/4	3/4	132
1×6	5 1/2	5 1/4	3/4	133
1 1/4×4	3 1/2	3 1/4	1 1/16	134
1 1/4×6	5 1/2	5 1/4	1 1/16	135
EASED EDGES				
3/4×4	3 1/2	3 1/4	11/16	632EE
3/4×6	5 1/2	5 1/4	11/16	633EE
1×4	3 1/2	3 1/4	3/4	132EE
1×6	5 1/2	5 1/4	3/4	133EE
V1S – 3/32 INCH V				
3/4×4	3 1/2	3 1/4	11/16	707
3/4×6	5 1/2	5 1/4	11/16	708
3/4×8	7 1/2	7 1/4	11/16	715
3/4×10	9 1/2	9 1/4	11/16	716
3/4×12	11 1/2	11 1/4	11/16	717
1×4	3 1/2	3 1/4	3/4	207
1×6	5 1/2	5 1/4	3/4	208
1×8	7 1/2	7 1/4	3/4	215
1×10	9 1/2	9 1/4	3/4	216
1×12	11 1/2	11 1/4	3/4	217

V1S – 1/4 INCH V

Size	B	A	C	No.
3/4×4	3 1/2	3 1/4	11/16	709
3/4×6	5 1/2	5 1/4	11/16	711
3/4×8	7 1/2	7 1/4	11/16	712
3/4×10	9 1/2	9 1/4	11/16	713
3/4×12	11 1/2	11 1/4	11/16	714
1×4	3 1/2	3 1/4	3/4	209
1×6	5 1/2	5 1/4	3/4	211
1×8	7 1/2	7 1/4	3/4	212
1×10	9 1/2	9 1/4	3/4	213
1×12	11 1/2	11 1/4	3/4	214

V2S – S1S – RESAWN 1S

Size	B	A	C	No.
3/4×4	3 1/2	3 1/4	11/16	709R
3/4×6	5 1/2	5 1/4	11/16	711R
3/4×8	7 1/2	7 1/4	11/16	712R
3/4×10	9 1/2	9 1/4	11/16	713R
3/4×12	11 1/2	11 1/4	11/16	714R
1×4	3 1/2	3 1/4	3/4	209R
1×6	5 1/2	5 1/4	3/4	211R
1×8	7 1/2	7 1/4	3/4	212R
1×10	9 1/2	9 1/4	3/4	213R
1×12	11 1/2	11 1/4	3/4	214R

*Figures in columns A, B, and C refer to dimensions as indicated in pattern drawings.

SIDING—REDWOOD BEVEL

Size (Nom.)	A*	B	C	Pattern No.
PLAIN BEVEL				
$1/2 \times 4$	$3 \, 1/2$	$2 \, 1/2$	$15/32 \times 3/16$	320
$1/2 \times 6$	$5 \, 1/2$	$4 \, 1/2$	$15/32 \times 3/16$	322
$1/2 \times 8$	$7 \, 1/2$	$6 \, 1/2$	$15/32 \times 3/16$	323
$5/8 \times 6$	$5 \, 1/2$	$4 \, 1/2$	$9/16 \times 3/16$	325
$5/8 \times 8$	$7 \, 1/2$	$6 \, 1/2$	$9/16 \times 3/16$	326
$5/8 \times 10$	$9 \, 1/2$	$8 \, 1/2$	$9/16 \times 3/16$	327
$3/4 \times 6$	$5 \, 1/2$	$4 \, 1/2$	$3/4 \times 3/16$	329
$3/4 \times 8$	$7 \, 1/2$	$6 \, 1/2$	$3/4 \times 3/16$	330
$3/4 \times 10$	$9 \, 1/2$	$8 \, 1/2$	$3/4 \times 3/16$	331
$3/4 \times 12$	$11 \, 1/2$	$10 \, 1/2$	$3/4 \times 3/16$	332
RABBETED BEVEL				
$1/2 \times 4$	$3 \, 1/2$	3	$1/2 \times 3/16$	360
$1/2 \times 6$	$5 \, 1/2$	5	$1/2 \times 3/16$	362
$1/2 \times 8$	$7 \, 1/2$	7	$1/2 \times 3/16$	363
$1/2 \times 4$	$3 \, 1/2$	3	$1/2 \times 3/16$	400†
$5/8 \times 4$	$3 \, 1/2$	3	$9/16 \times 3/16$	350
$5/8 \times 6$	$5 \, 1/2$	5	$9/16 \times 3/16$	352
$5/8 \times 8$	$7 \, 1/2$	7	$9/16 \times 3/16$	353
$3/4 \times 6$	$5 \, 1/2$	5	$11/16 \times 9/32$	371
$3/4 \times 8$	$7 \, 1/2$	7	$11/16 \times 9/32$	372
$3/4 \times 10$	$9 \, 1/2$	9	$11/16 \times 9/32$	373
$3/4 \times 12$	$11 \, 1/2$	11	$11/16 \times 9/32$	374
$3/4 \times 6$	$5 \, 1/2$	5	$3/4 \times 9/32$	391‡
$3/4 \times 8$	$7 \, 1/2$	7	$3/4 \times 9/32$	392‡
$3/4 \times 10$	$9 \, 1/2$	9	$3/4 \times 9/32$	393‡
$3/4 \times 12$	$11 \, 1/2$	11	$3/4 \times 9/32$	394‡

*Figures in columns A, B, and C refer to dimensions as indicated in pattern drawings.
†This pattern has round edge, $3/8$" radius.
‡Resawn face. These patterns have a square edge.

SIDING—VINYL

Type	Width (In.)	Length
Horiz.	8	10′ – 12′ 6″
Dbl. 4	8	10′ – 12′ 6″
Dbl. 6	12	10′ – 12′ 6″

Packed 1 square/ctn. Sizes, types may vary widely. Check suppliers for data. Vent and weep holes provided.

SIDING—WOOD

Type	Counted Width (In.)	Dressed Std. Thickness (In.)	Dressed Dim. Std. Face Width (In.)	Exp. Face (In.)	Add to Area
Bevel Siding	4	$7/16 \times 3/16$	3 1/2	2 3/4	45%
	4			3	33%
	5	$3/4 \times 3/16$	4 1/2	3 3/4	35%
	5			4	25%
	6	$3/4 \times 3/16$	5 1/2	4 1/2	33%
	6			4 3/4	27%
	6			5	20%
	8	$7/16 \times 3/16$	7 1/4	6	33%
	8			6 3/4	20%
	10	$9/16 \times 3/16$	9 1/4	7 3/4	29%
	10			8 3/4	15%
Drop and Rustic - (Shiplapped)	4	$9/16$	3 1/8		28%
	5	$3/4$	4 1/8		21%
	6		5 1/16		19%
	8		6 7/8		16%
Drop and Rustic (dressed &) (matched)	4	$9/16$	3 1/4		23%
	5	$3/4$	4 1/4		18%
	6		5 3/16		16%
	8		7		14%

SLAG—STONE (QUANTITIES/MILE)

Approximate Quantities of Slag and Stone, Including Screenings for Filler, per Mile of Pavement for Various Compact Depths and Various Widths.

Depth of Pavmt. Compact (In.)		Width of Pavement					
		10 ft.	12 ft.	14 ft.	16 ft.	18 ft.	20 ft.
		Tons of Aggregate/Mile					
4	Slag	1115	1338	1561	1784	2007	2230
4	Stone	1349	1619	1889	2159	2429	2699
6	Slag	1672	2006	2340	2674	3008	3342
6	Stone	2024	2429	2834	3238	3643	4048
8	Slag	2229	2675	3121	3566	4012	4458
8	Stone	2699	3239	3778	4318	4858	5398
10	Slag	2787	3344	3902	4459	5016	5574
10	Stone	3373	4048	4722	5397	6072	6747
12	Slag	3344	4013	4682	5350	6019	6688
12	Stone	4048	4858	5667	6477	7286	8096
14	Slag	3901	4681	5462	6242	7022	7803
14	Stone	4723	5667	6612	7556	8501	9445

Coarse aggregate, spread loose, will compact about 20% by rolling

SLAG—STONE (QUANTITIES/SQUARE YARD)
Quantities/Square Yard of Pavement for Various Compact Depths

Compact Depth (In.)	Slag			Stone		
	Tons/Square Yard			Tons/Square Yard		
	Coarse	Screenings	Total	Coarse	Screenings	Total
4	.1425	.0475	.1900	.1795	.0505	.2300
5	.1781	.0594	.2375	.2243	.0632	.2875
6	.2137	.0713	.2850	.2691	.0759	.3450
7	.2494	.0831	.3325	.3141	.0884	.4025
8	.2850	.0950	.3800	.3590	.1010	.4600
9	.3206	.1059	.4275	.4038	.1137	.5175
10	.3562	.1178	.4750	.4486	.1264	.5750
11	.3918	.1297	.5225	.4934	.1391	.6325
12	.4275	.1425	.5700	.5385	.1515	.6900
13	.4631	.1544	.6175	.5833	.1642	.7475

Slag screenings assumed to be 25% of the total weight.
Stone screenings assumed to be 25% of the total weight.

Soils—Types & Design Properties, see page 276.

SOLDER—COPPER PIPE

Quantity of Soft Solder Required to Make 100 Joints

Pipe Size (In.)	3/8	1/2	3/4	1	1 1/4	1 1/2	2	2 1/2	3	3 1/2	4	5	6	8	10
Quantity (Lbs.)	.5	.75	1.0	1.4	1.7	1.9	2.4	3.2	3.9	4.5	5.5	8.	15.	32.	40.

SOUND—ACOUSTICAL PROPERTIES

Coefficients of absorption for pitch of 512 vibrations per second:

Open Window 1.00
Brick wall, painted 017
Brick wall, set in portland cement .025
Cement 025
Concrete 015
Glass, single thickness 027
Marble 010
Plaster on wood lath 034
Plaster on metal lath 033
Plaster on tile. 025
Wood, plain 060
Wood, varnished 030

Sound, tables, see pages 279-281.
Spikes, tables, see page 281.

STAIRS—DISAPPEARING

Opening (In.)	Height	Width (In.)	Max. Proj. Into Room (In.)	Wt. (Lbs.)
26 x 54	7'-0" to 7'-6"	18	61	48
26 x 54	7'-7" to 8'-4"	18	63	50
26 x 54	8'-5" to 9'-0"	18	66	54
30 x 62	7'-9" to 8'-6"	22	76	80
30 x 62	8'-7" to 9'-3"	22	81	90
30 x 62	9'-4" to 10'-0"	22	86	100

Treads 4", Rise 9"
Proj. 12" into attic

SOILS—TYPES & DESIGN PROPERTIES

Soil Group	Unified Soil Classification Symbol	Soil Description	Allowable Bearing in Pounds/Square Foot with Medium Compaction or Stiffness[1]	Drainage Characteristics[2]	Frost Heave Potential	Volume Change Potential Expansion[3]
	GS	Well-graded gravels, gravel-sand mixtures, little or no fines.	8000	Good	Low	Low
	GP	Poorly graded gravels or gravel-sand mixtures, little or no fines.	8000	Good	Low	Low
Group I Excellent	SW	Well-graded sands, gravelly sands, little or no fines.	6000	Good	Low	Low
	SP	Poorly graded sands or gravelly sands, little or no fines.	5000	Good	Low	Low
	GM	Silty gravels, gravel-sand-silt mixtures.	4000	Good	Medium	Low
	SM	Silty sand, sand-silt mixtures.	4000	Good	Medium	Low

Group	Symbol	Description				
	GC	Clayey gravels, gravel-sand-clay mixtures.	4000	Medium	Medium	Low
Group II Fair to Good	SC	Clayey sands, sand-clay mixtures.	4000	Medium	Medium	Low
	ML	Inorganic silts and very fine sands, rock flour, silty or clayey fine sands or clayey silts with slight plasticity.	2000	Medium	High	Low
	CL	Inorganic clays of low to medium plasticity, gravelly clays, sandy clays, silty clays, lean clays.	2000	Medium	Medium	Medium[4]
	CH	Inorganic clays of high plasticity, fat clays.	2000	Poor	Medium	High[4]
Group III Poor	MH	Inorganic silts, micaceous or diatomaceous fine sandy or silty soils, elastic silts.	2000	Poor	High	High
	OL	Organic silts and organic silty clays of low plasticity	400	Poor	Medium	Medium
Group IV Unsatisfactory	OH	Organic clays of medium to high plasticity, organic silts.	—0—	Unsatisfactory	Medium	High
	Pt	Peat and other highly organic soils.	—0—	Unsatisfactory	Medium	High

[1] Allowable bearing value may be increased 25 percent for very compact, coarse grained gravelly or sandy soils, or very stiff fine-grained clayey or silty soils. Allowable bearing value shall be decreased 25 percent for loose, coarse-grained gravelly or sandy soils, or soft, fine-grained clayey or silty soils.

[2] The percolation rate for good drainage is over 4 inches per hour, medium drainage is 2 to 4 inches per hour, and poor is less than 2 inches per hour.

[3] For expansive soils, contact local soils engineer for verification of design assumptions.

[4] Dangerous expansion might occur if these soil types are dry but subject to future wetting.

STAIRS—RISERS
(WOOD)

Thick	Width	Lengths
3/4″	7 1/2″	3′, 3′ – 6″, 4′

STAIRS—SPIRAL (METAL)
Stair Table
Open Riser Design

Diameter	3′0	3′6	4′0	4′6	5′0	5′6	6′0	7′0	8′0
Treads/Circle	12	12	12	12	12	12	12	12	12
	16	16	16	16	16	16	16	16	16
Center Pole	3″	3″	3″	4″	4″	4″	4″	4″	4″

STAIRS—TREADS (NONSKID TOPPING)

- 2 Parts Portland Cement
- 1 Part Washed Sand
- 2 Parts 1/8 or 3/8 Aggregate
- 1 Part Carborundum Chips

STAIRS—TREADS (OAK)

Thick (In.)	Widths (In.)	Lengths
1 1/16	9 1/2	3′, 3′ – 6″, 4′
1 1/16	10 1/2	3′, 3′ – 6″; 4′
1 1/16	11 1/2	3′, 3′ – 6″, 4′

STAIRS—TYPICAL UNIT-RISE/UNIT-RUN COMBINATIONS

Floor-to-Floor Height	Unit Rise	Unit Run	Number of Risers	Number of Treads	Total Run
9'	7 3/4"	10 1/4"	14	13	11'
9'	7 1/4"	10 3/4"	15	14	12'6"
9'	6 3/4"	10 3/4"	16	15	13'
9'	6"	11"	18	17	15'8"
8'6"	7 1/4"	10 1/2"	14	13	11'4"
8'6"	7 1/4"	11 1/2"	14	13	12'6"
8'6"	6 3/8"	10 3/8"	16	15	13'
8'6"	6 3/8"	11"	16	15	13'9"
8'	7 3/8"	10"	13	12	10'
8'	7 3/8"	11"	13	12	11'
8'	6"	10 3/8"	16	15	13'
8'	6"	12"	16	15	15'

STAPLES

Length (In.)	Ga.	No./ Lb.
3/4	14	480
7/8	14	416
1	9	108
1 1/4	9	87
1 1/2	9	72
1 3/4	9	65

SOUND—HEARING RATINGS

Loss in Decibels	Rating	Condition
Under 30	Bad	Normal speech heard through walls
30 to 35	Fair	Loud speech heard through walls
35 to 40	Good	Normal speech not heard
40 to 45	Very Good	Loud speech faintly heard
Over 45	Excellent	Loud sounds heard faintly

SOUND—IRRITATION LEVELS

Sound or Noise	Approx. Range Decibel Levels. Ratings, Normal Conditions
Air Conditioners, Window	70 to 90
Alarm Clocks	75 to 95
Birds, Chirping Av.	50 to 70
Chain Saws	90 to 110
Parties, Cocktail, Large	65 to 85
Diesel Trucks, Large	70 to 80
Faucets Dripping	40 to 50
Jackhammers	90 to 110
Jet Planes, Takeoffs	125 to 150
Kitchen Mixers	65 to 75
Motor Cycles	90 to 110
Mowers, Power	95 to 115
Rainfall, Moderate	40 to 50
Refrigerators, Freezers	40 to 60
Shaver, Electric	75 to 85
Sports Crowds	100 to 125
Thunder, Loud	110 to 130
Traffic, Normal	70 to 90
Vacuum Sweepers, Home	65 to 80
Washing Machines, Home	60 to 75

Proper sound insulation may be used to control irritating noises in inhabited areas.

SOUND—NOISE LEVELS

Condition	Av. Noise Level Decibels
Average Residence	
Without Radio	43.0
With Radio	50.0
Small Store (1-5 Clerks)	53.5
Large Store (more than 5 Clerks)	61.0
Small Office (1 or 2 Desks)	53.5
Medium Office (3-10 Desks)	58.0
Large Office (more than 10 Desks)	64.5
Factory Office	61.5
Miscellaneous Business	56.0
Factory	77.0

SOUND—TRANSMISSION VALUES

Type Wall		Loss in Decibels
4″ Lt. Wt. Conc. Block	Unplastered	.37
4″ Lt. Wt. Conc. Block	Plastered	.43
6″ Lt. Wt. Conc. Block	Unplastered	.45
6″ Lt. Wt. Conc. Block	Plastered	.49
8″ Lt. Wt. Conc. Block	Unplastered	.48
8″ Lt. Wt. Conc. Block	Plastered	.51
12″ Lt. Wt. Conc. Block	Unplastered	.52
12″ Lt. Wt. Conc. Block	Plastered	.54
2×4 Studs Gyp. Lath	Plaster	.411
2×4 Studs Insul. Lath	Plaster	.409
2×4 Studs 1/4″ Plywood	Both Sides	.31
1/4″ Plywood		.224
1/2″ Fiberboard		.223

SPIKES—COMMON WIRE

Size	Length (In.)	Diameter (In.)	Size	Length (In.)	Diameter (In.)
10d	3	0.192	40d	5	0.263
12d	3 1/4	.192	50d	5 1/2	.283
16d	3 1/2	.207	60d	6	.283
20d	4	.225	5/16 inch	7	.312
30d	4 1/2	.244	3/8 inch	8 1/2	.375

SPIKES—GUTTER

Length (In.)	Gauge (In.)	Head Size (In.)	No./ Lb.
6	3/16	3/8	24
7	3/16	3/8	17
8	3/16	3/8	15
8	1/4	9/16	9
9	1/4	9/16	8
10	1/4	9/16	7

SUMP—CAPACITIES

Diameter (Feet)	Gals./ Foot Depth
1 1/2	15
2	25
2 1/2	40
3	50
3 1/2	75
4	95
5	150
6	210

STUCCO—COVERAGE PER BARREL OF CEMENT
(VARIOUS MIXES)

Mix Parts by Volume		Thickness of Coat			
Cement	Sand	1/4 In./Sq. Ft.	3/8 In./Sq. Ft.	1/2 In./Sq. Ft.	3/4 In./Sq. Ft.
1	1	266	177	133	89
1	1 1/2	336	226	168	112
1	2	404	270	202	135
1	2 1/2	472	314	236	157
1	3	542	362	271	181
1	3 1/2	612	408	306	204
1	4	682	455	341	227

STUCCO—MATERIALS FOR 100 SQUARE FEET
(ROUGH SURFACE)

Thick-ness (In.)	1:1 Mix		1:2 Mix		1:2 1/2 Mix		1:3 Mix	
	Cement Sacks	Sand Cu. Yd.	Cement Sacks	Sand Cu. Yd.	Cement Sacks	Sand Cu. Yd.	Cement Sacks	Sand Cu. Yd.
3/8	2.5	0.08	1.5	0.11	1.3	0.12	1.1	0.13
1/2	3.0	0.11	2.0	0.15	1.7	0.16	1.5	0.17
3/4	4.5	0.16	3.0	0.22	2.5	0.23	2.2	0.25
1	6.0	0.22	4.0	0.29	3.3	0.31	3.0	0.33
1 1/4	7.5	0.27	5.0	0.36	4.2	0.39	3.7	0.41
1 1/2	9.0	0.33	6.0	0.43	5.1	0.47	4.5	0.50
1 3/4	10.5	0.39	7.0	0.50	6.0	0.56	5.4	0.60
2	12.0	0.45	8.0	0.58	6.9	0.64	6.2	0.69

STUCCO—MATERIALS FOR 100 SQUARE FEET
(SMOOTH SURFACE)

Thick-ness (In.)	Cu. Ft. Mortar	1:2 Mix		1:2 1/2 Mix		1:3 Mix		1:3 1/2 Mix	
		Cement Sacks	Sand Cu. Ft.	Cement Sacks	Sand Cu. Ft.	Cement Sacks	Sand Cu. Ft.	Cement Sacks	Sand Cu. Ft.
1/4	2.08	.93	1.86	.79	1.96	.68	2.06	.60	2.10
3/8	3.13	1.40	2.80	1.19	2.95	1.03	3.10	.90	3.16
1/2	4.17	1.86	3.72	1.58	3.94	1.37	4.12	1.20	4.22
5/8	5.21	2.32	4.65	1.98	4.92	1.71	5.15	1.50	5.27
3/4	6.25	2.79	5.58	2.37	5.90	2.06	6.18	1.80	6.32
1	8.33	3.72	7.44	3.16	7.88	2.74	8.24	2.40	8.42

Sump—Capacities, see page 281.

T

TACKS—SIZES, LENGTHS

Size No.	1	1 1/2	2	2 1/2	3	4	6
Cut (In.).	3/16	7/32	1/4	5/16	3/8	7/16	1/2
Carpet (In.)	3/16	7/32	1/4	5/16	3/8	7/16	1/2
Gimp (In.)	7/32	1/4	5/16	3/8	7/16	1/2	9/16
Upholsterers (In.)	3/16	7/32	1/4	3/16	3/8	7/16	1/2
Size No.	**8**	**10**	**12**	**14**	**16**	**18**	**20**
Cut (In.).	9/16	5/8	11/16	3/4	13/16	7/8	15/16
Carpet (In.)	9/16	5/8	11/16	3/4	13/16	7/8	15/16
Gimp (In.)	5/8	11/16	3/4	13/16	7/8	15/16	1
Upholsterers (In.)	9/16	5/8	11/16	3/4	13/16	7/8	15/16

TANKS—CONCRETE (MATERIALS REQUIRED)

Size—Dia. × Height (Ft.)	Cement Bbls.	Sand (Cu. Yds.)	Stone (Cu. Yds.)
10 × 25	25.0	7.5	15.0
12 × 30	36.0	11.0	22.0
12 × 35	41.0	12.5	25.0
14 × 30	42.5	13.0	26.0
14 × 35	48.0	14.7	29.4
14 × 40	54.5	16.5	33.0
16 × 40	62.5	19.0	38.0
16 × 45	69.0	21.0	42.0
18 × 45	80.0	24.5	49.0
18 × 50	86.5	26.5	53.0

Wall thickness: 6 inches

TANKS—OIL

Size	Ga.	Capacity	Opngs.
OVAL TYPE			
22 × 44 × 60	12	235	4 – 2″
27 × 44 × 48	12	220	4 – 2″
27 × 44 × 60	12	275	4 – 2″
ROUND TYPE			
38 × 60	12	285	4 – 2″
46 × 76	12	550	4 – 2″
52 × 110	10	1000	4 – 2″

Other sizes available

THRESHOLDS—METAL

Type	Thick (In.)	Width (In.)	Lengths
Plain Top	$1/4$	2 $1/2$ to 6	2′7″, 2′9″, 3′1″
Fluted Top	$1/2$	4, 5, 6	2′7″, 2′9″, 3′1″

Numerous styles available

THRESHOLDS—OAK

Thick (In.)	Width (In.)	Lengths
$5/8$	3 $5/8$	2′7″, 2′9″, 3′1″
$3/4$	3 $5/8$	2′7″, 2′9″, 3′1″
1 $1/16$	3 $5/8$	2′7″, 2′9″, 3′1″
$3/4$	4 $1/2$	2′7″, 2′9″, 3′1″
$3/4$	5 $1/2$	2′7″, 2′9″, 3′1″

Other sizes available

TIES—REINFORCING BAR
Packed 5,000 Ties/Bundle, 10-Gauge Wire Lengths

No.	Bar Sizes (In.)	1/4"	3/8"	1/2"	5/8"	3/4"	7/8"	1"	1 1/8"	1 1/4"
		Length in Inches of Tie Required – Round or Square Bars								
2	1/4	3	3 1/2	4	4 1/2	5	5 1/2	6	6 1/2	7
3	3/8	3 1/2	4	4 1/2	5	5 1/2	6	6 1/2	7	7 1/2
4	1/2	4	4 1/2	5	5	5 1/2	6	6 1/2	7 1/2	8
5	5/8	4 1/2	5	5	5 1/2	6	6 1/2	7	7 1/2	8
6	3/4	5	5 1/2	5 1/2	6	6 1/2	6 1/2	7 1/2	8	8 1/2
7	7/8	5 1/2	6	6	6 1/2	6 1/2	7	7 1/2	8 1/2	8 1/2
8 – 9	1	6	6 1/2	7	7	7 1/2	8	8 1/2	9	9
10	1 1/8	6 1/2	7	7 1/2	7 1/2	8	8 1/2	9	9	10
11	1 1/4	7	7 1/2	8	8	8 1/2	8 1/2	9	10	10

TILE—CERAMIC (ESTIMATING DATA, FLOOR & WALL TYPES)

Dimensions, (In.)		Per Sq. Ft.			Packaging (Per Carton)		
Nominal	Actual	Thickness (In.)	Weight (Lbs.)	Pieces Req'd.	Pieces	Sq. Ft.	Weight (Lbs.)
4 × 4	3 5/16 × 3 5/16	3/8	3.8	9.28	64	8.40	40
4 × 8	3 5/16 × 7 7/8	3/8	3.8	4.64	50	10.78	40
6 × 6	5 5/16 Sq.	1/4	2.7	4.13	63	15.26	40
8 × 8	7 7/8 Sq.	1/4	3.8	2.32	30	12.93	50
8 × 12	7 7/8 × 11 5/8	3/8	4.5	1.49	16	10.73	47
12 × 12	11 3/4 Sq.	3/8	4.4	1.02	15	14.53	64
12 1/2 × 12 1/2	12 3/8 Sq.	3/8	4.5	.93	14	14.96	60
12 × 16	11 13/16 × 15 5/8	3/8	4.6	.75	10	13.33	60
12 × 24	11 13/16 × 23 5/8	3/8	4.6	.52	8	15.50	70
24 × 24	23 5/8 Sq.	1/2	5.9	.26	6	23.25	137

6" & 8" Hexagonal & 8" Octagonal may be available.

Check suppliers for various types, methods and quantities of setting products, grouting materials, unusual shapes, sizes may vary.

Trim Shapes: Bullnose, Bullnose Corners, Cove Base, Cove Base Out Corner, Internal Corner Cove Base, Cove Base Stop, External and Internal Corners, Cove Cap Connectors, Raised Cap Corners, Stair Treads.

Some Types have self spacing lugs.

TILE—CERAMIC (METRIC CONVERSIONS)

Nominal Size (Inches)	Nominal Size (Centimeters)	Actual Size (Centimeters)
2 × 6 (on 6 × 6 sheets)	5 × 15	4.5 × 4.5
3 × 3 (12 × 12 sheets)	7.5 × 7.5	7 × 7
4 × 4	10 × 10	9 × 9
4 × 8	10 × 20	9 × 19
4 3/4 × 4 3/4	12 × 12	11 × 11
6 × 6	15 × 15	14 × 14
7 1/2 Hex.	19 Hex.	18 Hex.
7 × 7	18 × 18	17 × 17
8 × 8	20 × 20	19 × 19
9 1/2 × 9 1/2	24 × 24	23 × 23
12 × 12	30 × 30	29 × 29

TILE—COMBUSTIBLE ACOUSTICAL (FIBER INSULATING PERFORMANCE)

Thickness (In.)	Sizes	Wt. Sq. Ft.	K Factor	Noise Red.
1/2	12 × 12 – 12 × 24	.75	.38	.60
3/4	12 × 12 – 12 × 24	.85	.38	.65
1	12 × 12 – 12 × 24	1.25	.38	.70

Tile—Floor, see page 287.

TILE—QUARRY

Types	Sizes (In.)	Thickness (In.)
Glazed Int.	4 1/4 × 4 1/2 – 6 × 6	3/8
Extra Duty	4 1/4 × 4 1/2 – 6 × 6	3/8 to 5/8
Faience	4 1/4 × 4 1/2 – 6 × 6	1/2 to 5/8
Paver	3 × 3 – 4 × 4 – 4 1/4 × 4 1/4 – 6 × 6 – 6 × 3	3/8 to 5/8
Quarry	2 3/4 × 2 3/4 – 4 × 4 – 6 × 6 – 8 × 8 – 9 × 9 – 6 × 2 3/4 – 6 × 9 – 8 × 3 3/4	1/2 to 3/4
	8 × 4	1 1/4 to 1 5/8
Ship	6 × 6	5/8 to 3/4

TILE—FLOOR (QUANTITIES)

Square Feet	Number of Tile Needed		
	9″ × 9″	12″ × 12″	9″ × 18″
1	2	1	1
2	4	2	2
3	6	3	3
4	8	4	4
5	9	5	5
6	11	6	6
7	13	7	7
8	15	8	8
9	16	9	8
10	18	10	9
20	36	20	18
30	54	30	27
40	72	40	36
50	89	50	45
60	107	60	54
70	125	70	63
80	143	80	72
90	160	90	80
100	178	100	90
200	356	200	178
300	534	300	267
400	712	400	356
500	890	500	445
600	1068	600	534
700	1246	700	623
800	1424	800	712
900	1602	900	801
1000	1780	1000	890
Labor/ 100 Sq. Ft.	2 Hrs.	1.3 Hrs.	1.3 Hrs.

Tile Waste Allowance	1 to 50 sq. ft. 14%
	50 to 100 sq. ft. 10%
	100 to 200 sq. ft. 8%
	200 to 300 sq. ft. 7%
	300 to 1000 sq. ft. 5%
	Over 1000 sq. ft. 3%

TILE—TYPES & SIZES

Types	Sizes (In.)
Dots	1 1/2, 11/32
Squares	3/4, 11/16, 1 9/16, 2, 4, 4 1/4, 6, 8, 12
Rectangles	3/4 × 1 9/16, 1/2 × 1 1/16, 1 × 2, 11/16 × 3/16, 4 1/4 × 6 4 × 8, 6 × 9, 6 × 12
Hexagonal	1, 1 1/4, 2, 3, 4, 6, 8, 12
Octagonal	2 3/16, 4, 6, 8, 10, 12

Pentagonal available in some sizes: 1/4, 5/16, 1/2, 3/4, 7/8 in.
Self-spaced glued-on or ready-to-apply sheets of small tile up to 4 × 4 in. may be available in slabs 12 × 12 and 12 × 24 in.

TILE—VINYL

- Package: 45-Sq. Ft. Cartons
- Size: 12 × 12 Inches (self-adhesive)
- Use: Household, Miscellaneous
- Thickness: 1.2 mm, 1.3 mm, 1.5 mm
- Commercial Type: 12 × 12 Inches, 1/8 Inch Thick

TILE—VITREOUS CERAMIC, 1/4 INCH (TYPICAL SIZES)

Type	Sizes (In.)
Squares	1/2 × 1/2 3/4 × 3/4 – (1 × 1 to 9 × 9, 5/8 thick) 1 1/16 × 1 1/16 1 1/32 × 1 1/32 1 9/16 × 1 9/16 2 1/8 × 2 1/8 2 3/16 × 2 3/16
Oblong	1/2 × 1 1/16 to 1/2 × 6 1 × 6 – 2 × 6 – 3 × 6 3/4 × 1 9/16 1 1/32 × 2 1/8 1 1/16 × 2 3/16
Hexagonal	1, 1 1/4, 2
Octagonal	1, 2, 2 3/16
Pentagon	2, 2 3/16
Triangular	1 1/32 × 1 1/32 2 1/8 × 2 1/8

TIMBERS—FIR

Size (In.)	Wt.— Green, (Lbs.)
Rough	3500
6× 6 S4S	3000
6× 8 S4S	3100
6×10 S4S	3100
6×12 S4S	3100
8× 8 S4S	3100
8×10 S4S	3200
8×12 S4S	3200
10×10 S4S	3200
10×12 S4S	3200
12×12 S4S	3300

Timbers—Sizes, Areas, Weights, see pages 290-291.

TRUSSES—ROOF (WOOD)
Approximate Lumber/Truss
3 Span Residence Types for Plastered Ceilings

Roof Slopes	Truss Lengths						
	20′	22′	24′	26′	28′	30′	32′
4 – 12	50	60	60	70	90	100	100
5 – 12	55	60	65	70	80	90	95
6 – 12	55	65	70	70	90	95	100
7 – 12	60	65	70	75	90	100	100

Deduct 10% for dry wall ceilings all 28′, 30′, 32′ trusses

TIMBERS—SIZES, AREAS, WEIGHTS

Size (In.)	Dressed Size (In.)	Area of Section (Sq. In.)	Approx. Wt./Foot (Lb.)
2× 4	1 ½ × 3 ½	5.25	1.64
6	5 ½	8.25	2.54
8	7 ¼	10.9	3.39
10	9 ¼	13.9	4.29
12	11 ¼	16.9	5.19
14	13 ¼	19.9	6.09
16	15 ¼	22.9	6.99
18	17 ¼	25.9	7.90
3× 4	2 ½ × 3 ½	8.75	2.64
6	5 ½	13.8	4.10
8	7 ¼	18.1	5.47
10	9 ¼	23.1	6.93
12	11 ¼	28.1	8.39
14	13 ¼	33.1	9.84
16	15 ¼	38.1	11.3
18	17 ¼	43.1	12.8
4× 4	3 ½ × 3 ½	12.3	3.65
6	5 ½	19.3	5.66
8	7 ½	27.2	7.55
10	9 ½	34.4	9.57
12	11 ½	41.7	11.6
14	13 ½	48.9	13.6
16	15 ½	56.2	15.6
18	17 ½	63.4	17.6
6× 6	5 ½ × 5 ½	30.3	8.40
8	7 ½	41.3	11.4
10	9 ½	52.3	14.5
12	11 ½	63.3	17.5
14	13 ½	74.3	20.6
16	15 ½	85.3	23.6
18	17 ½	96.3	26.7
20	19 ½	107.3	29.8
8× 8	7 ½ × 7 ½	56.3	15.6
10	9 ½	71.3	19.8
12	11 ½	86.3	23.9
14	13 ½	101.3	28.0

16	15 1/2	116.3	32.0
18	17 1/2	131.3	36.4
20	19 1/2	146.3	40.6
22	21 1/2	161.3	44.8
10×10	9 1/2× 9 1/2	90.3	25.0
12	11 1/2	109	30.3
14	13 1/2	128	35.6
16	15 1/2	147	40.9
18	17 1/2	166	46.1
20	19 1/2	185	51.4
22	21 1/2	204	56.7
24	23 1/2	223	62.0
12×12	11 1/2×11 1/2	132	36.7
14	13 1/2	155	43.1
16	15 1/2	178	49.5
18	17 1/2	201	55.9
20	19 1/2	224	62.3
22	21 1/2	247	68.7
24	23 1/2	270	75.0
14×14	13 1/2×13 1/2	182	50.6
16	15 1/2	209	58.1
18	17 1/2	236	65.6
20	19 1/2	263	73.1
22	21 1/2	290	80.6
24	23 1/2	317	88.1
16×16	15 1/2×15 1/2	240	66.7
18	17 1/2	271	75.3
20	19 1/2	302	83.9
22	21 1/2	333	92.5
24	23 1/2	364	101
18×18	17 1/2×17 1/2	306	85.0
20	19 1/2	341	94.8
22	21 1/2	376	105
24	23 1/2	411	114
26	25 1/2	446	124
20×20	19 1/2×19 1/2	380	106
22	21 1/2	419	116
24	23 1/2	458	127
26	25 1/2	497	138
28	27 1/2	536	149
24×24	23 1/2×23 1/2	552	153
26	25 1/2	599	166
28	27 1/2	646	180
30	29 1/2	693	193

U

U-VALUES

PLAIN UNITS

CORES OF UNITS
(Filled with Expanded Perlite or Vermiculite)

| Wall Type | Interior Finish | | | Interior Finish | | |
| | | Insulating Plastic Over | | | Insulating Plastic Over | |
	Plain Wall	Wall Direct 5/8" Thick	Furring, 3/8" Gypsum Lath, 1/2" Plaster	Plain Wall	Wall Direct 5/8" Thick	Furring, 3/8" Gypsum Lath, 1/2" Plaster
8" Concrete Block Masonry Walls— Aggregate:						
Haydite	.35	.30	.22	.18	.17	.14
Cinder	.39	.34	.24	.20	.19	.15
Sand & Gravel	.52	.43	.28	.38	.33	.17
10" Cavity Wall Face Brick, 2 1/2" Air Space, Common Brick	.33	.29	.22	.14	.13	.12
10" Cavity Wall 4" Common Brick, 2" Air Space, 4" Concrete Block (Gravel Agg.)	.30	.27	.20	.15	.15	.12

U-VALUES—CEILINGS

Ceilings	U
Ceilings joists, gyp. wallboard................................. full 6″ thickness wool insulation over joists.	.047
Ceiling joists, gyp. wallboard.................................. full 4″ thickness wool insulation over joists.	.0631
Ceiling joists, gyp. wallboard.................................. 3″ thickness wool insulation between joists.	.080
Ceiling joists, gyp. wallboard.................................. full thick batt insulation between joists.	.080
Ceiling joists, gyp. wallboard.................................. 2″ blanket insulation between joists.	.126
Ceiling joists, gyp. wallboard.................................. 1″ blanket insulation between joists.	.192
Ceiling joists, gyp. wallboard.................................. reflective paper and air space.	.207
Ceiling joists, gyp. wallboard.................................. no flooring, not insulated.	.621

U-VALUES—CONCRETE MASONRY
Typical Insulation Value—Expanded Perlite or Vermiculite

Block		"U" Factor	
		Without Fill	With Fill
8 × 8 × 16	Lt. Weight	0.33	0.152
8 × 8 × 16	Std. Weight	0.53	0.338
8 × 12 × 16	Lt. Weight	0.48	0.113
8 × 12 × 16	Std. Weight	0.30	0.221

U-VALUES—CONCRETE MASONRY WALLS

Construction	Interior Finish			
	Plain Wall No Plaster	Wall Direct	1/2-in. Plaster over: 1/4-in. Furring with:	
			3/8-In. Plaster Board	1/2-In. Rigid Insulation
CONCRETE MASONRY				
8-in. sand and gravel or limestone	0.52	0.49	0.31	0.22
8-in. cinder	0.37	0.35	0.25	0.19
8-in. expanded slag, clay or shale	0.33	0.32	0.24	0.18
12-in. sand and gravel or limestone	0.49	0.45	0.30	0.22
12-in cinder	0.35	0.33	0.24	0.18
12-in. expanded slag, clay or shale	0.32	0.31	0.23	0.18
CONCRETE MASONRY (Cores Filled with Vermiculite or Perlite)				
8-in. sand and gravel or limestone	0.39	0.38	0.26	0.19
8-in. cinder	0.20	0.19	0.16	0.13
8-in. expanded slag, clay or shale	0.17	0.17	0.14	0.12
12-in. sand and gravel or limestone	0.34	0.32	0.24	0.18
12-in. cinder	0.20	0.19	0.15	0.13
12-in. expanded slag, clay or shale	0.15	0.14	0.12	0.11
4-INCH FACE BRICK PLUS:				
4-in. sand and gravel or limestone unit	0.53	0.49	0.31	0.23
4-in. cinder, expanded slag, clay or shale unit	0.44	0.42	0.28	0.21
4-in. common brick	0.50	0.46	0.30	0.22
8-in. sand and gravel or limestone unit	0.44	0.41	0.28	0.21
8-in. cinder, expanded slag, clay or shale unit	0.31	0.30	0.22	0.17
8-in. common brick	0.36	0.34	0.24	0.19
1-in. wood sheathing, paper, 2 × 4 studs, wood lath and plaster	—	0.27	0.27	0.20

U-VALUES—MISCELLANEOUS SECTIONS

Type Construction	Typical U Values
Brick	
4″ Face Brick & 4″ Common Brick	.50
4″ Face Brick & 8″ Common Brick	.36
4″ Face Brick & 12″ Common Brick	.29
Brick & Hollow Tile	
4″ Face Brick & 6″ Hollow Tile	.36
4″ Face Brick & 8″ Hollow Tile	.35
4″ Face Brick & 10″ Hollow Tile	.34
4″ Face Brick & 12″ Hollow Tile	.27
4″ Face Brick, 4″ Common Brick & 6″ Hollow Tile	.28
4″ Face Brick, 4″ Common Brick & 8″ Hollow Tile	.27
Brick & Concrete Block	
4″ Face Brick & 8″ Block (Gravel Aggregate)	.45
4″ Face Brick & 12″ Block (Gravel Aggregate)	.41
4″ Face Brick & 8″ Block (Cinder Aggregate)	.35
4″ Face Brick & 8″ Block (Lightweight Aggregate)	.31
Brick & Poured Concrete	
4″ Face Brick & 6″ Concrete (Gravel Aggregate)	.59
4″ Face Brick & 8″ Concrete (Gravel Aggregate)	.54
4″ Face Brick & 10″ Concrete (Gravel Aggregate)	.50
4″ Face Brick & 12″ Concrete (Gravel Aggregate)	.46
Poured Concrete	
6″ Concrete	.79
8″ Concrete	.70
10″ Concrete	.63
12″ Concrete	.57
Concrete Block	
8″ Block (Gravel Aggregate)	.56
12″ Block (Gravel Aggregate)	.49
Cut Stone & Hollow Tile	
4″ Stone & 6″ Hollow Tile	.37
4″ Stone & 8″ Hollow Tile	.36
Cut Stone & Concrete Block	
4″ Stone & 8″ Block (Gravel Aggregate)	.48
4″ Stone & 8″ Block (Cinder Aggregate)	.36
4″ Stone & 8″ Block (Lightweight Aggregate)	.32
Cut Stone & Poured Concrete	
4″ Stone & 6″ Concrete	.63
4″ Stone & 8″ Concrete	.57
4″ Stone & 10″ Concrete	.53
4″ Stone & 12″ Concrete	.49

Type Construction	Typical U Values
Wood Siding 1/2″ Gyp. Shtg. 3/4″ Gyp. Lath & Pl.	.31
Wood Siding 1/2″ Gyp. Shtg. 3/4″ Insul. Lath & Pl.	.22
Wood Shgls 1/2″ Gyp. Shtg. 3/4″ Gyp. Lath & Pl.	.24
Wood Shgls 1/2″ Gyp. Shtg. 3/4″ Insul. Lath & Pl.	.19
Stucco 3/4″ Gyp. Lath & Pl.	.39
Stucco 3/4″ Lath & Pl.	
Brick Veneer 1/2″ Gyp. Shtg. 3/4″ Gyp. Lath & Pl.	.34
8″ Brick Wall Furred 3/4″ Gyp. Lath & Pl.	.30

U-VALUES—ROOF CONSTRUCTION

Roof Deck Construction	Roofing Matr'l	Thickness (In.)	U Factor
Steel deck with 1″ Insulation Board on top of deck	3/8″ Built up	18 ga, 20 ga. and 22 ga.	0.25
Precast Reinforced Lightweight concrete tile	3/8″ Built up	2 3/4 3 1/2	0.52 0.49
Reinforced concrete slab	3/8″ Built up	2 4 6	0.82 0.72 0.64
Wood Deck	3/8″ Built up	1 2	0.49 0.32
2″ or 2 1/2″ gypsum poured on 1/2″ gypsum board form	3/8″ Built up	2 1/2 3	0.38 0.34
Reinforced gypsum plank - 2″ thick - (metal edged)	3/8″ Built up	2	0.58
Precast reinforced gypsum tile (3″ solid short span)	3/8″ Built up	3	0.48
Laminated gypsum board 2, 3 or 4 1/2″ plies	3/8″ Built up	1 1 1/2 2	0.57 0.47 0.40

U-VALUES—WALL CONSTRUCTION

Exterior Const.	Type of Interior Finish and Insulation	U Factor
SCR Brick - 2 × 2 Furring	1″ roll insul, 1/2″ insul board lath, 1/2″ vermiculite or perlite pl.	0.12
"	1″ roll insul, 2/8″ gyp lath, 3/4″ vermiculite or perlite pl.	0.14
"	1″ roll insul, metal lath, 3/4″ vermiculite or perlite pl.	0.15
"	1″ roll insul, 3/8″ gyp bd (dry wall)	0.16
"	1″ roll insul, metal lath, 3/4″ gyp pl.	0.16
"	1″ roll insul, 3/8″ gyp lath, 1/2″ gyp pl.	0.16
"	3/8″ gyp lath with al foil, 1/2″ vermiculite or perlite pl.	0.23
"	1/2″ Insul bd lath, 1/2″ vermiculite or perlite pl.	0.23
"	1/2″ insul bd lath, 1/2″ gyp pl.	0.25
"	3/8″ gyp lath with al foil, 1/2″ gyp pl.	0.25
"	1/2″ gyp bd (dry wall) with al foil	0.26
"	Metal lath, 3/4″ vermiculite or perlite pl.	0.33
"	3/8″ gyp lath, 1/2″ vermiculite or perlite pl.	0.33
"	3/8″ gyp lath, 1/2″ gyp pl.	0.37
"	Metal lath, 3/4″ gyp pl.	0.40
3/8″ Ext. Plywood	1/4″ Plywood	0.40
" "	3/8″ Plywood	0.38
" "	1/4″ Plywood, 1/2″ Blanket Insul	0.19
5/16″ Ply furring, siding	1/4″ Plywood	0.23

5/16" Ply furring, 3/8" Ext. Plywood	1/4" Plywood	0.26
5/16" Ply furring, 3/8" Ext. Plywood	1/4" Plywood, 1/2" Blanket Insul.	0.15
5/16" Ply Paper, siding	3/4" Wood lath and plaster	0.29
5/16" Ply Paper, siding	3/4" Wood lath and plaster, 1/2" Bl. Insul	0.16
8" Brick, 3/4" Furring	3/8" gyp lath, 1/2" pl	0.28
"	3/8" insul gyp lath, 1/2" pl	0.22
Wood Sdg, 1/2" gyp shtg.	3/8" insul gyp lath, 1/2" pl	0.23
Brick Veneer, "	3/8" gyp lath, 1/2" pl	0.31
"	3/8" gyp lath, 1/2" pl	0.34
Stucco "	3/8" insul gyp lath, 1/2" pl	0.23
"	3/8" insul gyp lath, 1/2" pl	0.25
Wood Shgls "	3/8" gyp lath, 1/2" pl	0.39
"	3/8" gyp lath, 1/2" pl	0.30
Asbestos Shgls "	3/8" insul gyp lath, 1/2" pl	0.22
"	3/8" insul gyp lath, 1/2" pl	0.26
8" Conc. Masonry	3/8" gyp lath, 1/2" pl	0.38
8" "	Plaster on walls, 1/2" pl.	0.53
8" "	Insul lath & plaster furred, 1/2" pl	0.22
12" "	Gyp lath & plaster, 1/2" pl	0.31
12" "	Gyp lath & plaster, 1/2" pl	0.30
12" "	Insul lath & plaster furred, 1/2" pl	0.22
Wood Sdg.	Plaster on walls, 1/2" pl.	0.49
	Gyp wallboard.	0.61

V

VAULT CONSTRUCTION—WALL SIZES

Kind of Material	Reinforced Concrete			Brick			Hollow Units
Class of Wall	6 Hr.	4 Hr.	2 Hr.	6 Hr.	4 Hr.	2 Hr.	2 Hr.
Floor – Counting from Top Down	Thickness of Wall (In.)						
Top	10	8	6	12	12	8	8
2nd from top	10	8	8	12	12	12	12
3rd from top	10	10	10	12	12	12	12
4th from top	12	10	10	16	16	16	16
5th from top	12	12	12	16	16	16	16
6th from top	12	12	12	20	16	16	16
7th from top	12	12	12	16	16	16	16
8th from top	12	12	12	16	16	16	16
9th from top	12	12	12	16	16	16	16
10th from top	14	12	12	16	16	16	16

VENTILATION—ATTIC

Width (in Feet)	20	22	24	26	28	30	32	34	36	38	40	42
20	192	211	230	250	269	288	307	326	346	365	384	403
22	211	232	253	275	296	317	338	359	380	401	422	444
24	230	253	276	300	323	346	369	392	415	438	461	484
26	250	275	300	324	349	374	399	424	449	474	499	524
28	269	296	323	349	376	403	430	457	484	511	538	564
30	288	317	346	374	403	432	461	490	518	547	576	605
32	307	338	369	399	430	461	492	522	553	584	614	645
34	326	359	392	424	457	490	522	555	588	620	653	685
36	346	380	415	449	484	518	553	588	622	657	691	726
38	365	401	438	474	511	547	584	620	657	693	730	766
40	384	422	461	499	538	576	614	653	691	730	768	806
42	403	444	484	524	564	605	645	685	726	766	806	847
44	422	465	507	549	591	634	676	718	760	803	845	887
46	442	486	530	574	618	662	707	751	795	839	883	927
48	461	507	553	599	645	691	737	783	829	876	922	968
50	480	528	576	624	672	720	768	816	864	912	960	1008

Sq. in. of net vent area req'd for attic area.
Roofs over 2″ in 12″ slope (Less than 2″ in 12″ slope double vent area.)
Use outside length and width dimensions.

Length (in Feet)

Based on ratio of minimum net ventilator area to ceiling area of 1/300

VENTILATORS—BRICK AND BLOCK

Type	Size (In.)	Net Free Vent Area (Sq. In.)
Standard Brick and Block	2 1/2 × 8	5.88
	2 1/2 × 16 3/8	11.76
	5 × 8	17.20
	5 × 16 3/8	34.40
	8 1/16 × 8	30.10
	8 1/16 × 16 3/8	60.20
Modular Brick and Block	2 1/2 × 7 3/4	5.88
	2 1/2 × 15 5/8	11.76
	5 × 7 3/4	17.20
	5 × 15 5/8	34.40
	8 1/6 × 7 3/4	30.10
	8 1/16 × 15 5/8	60.20

Numerous other sizes available

VENTILATORS—FOUNDATION

Size	Type	Aluminum	Wt.	Cast Iron	Wt.
8 × 8	Bar-Diagonal	x	1 1/4	x	3
12 × 8	Bar-Diagonal	x	1 3/4	x	4
16 × 8	Bar-Diagonal	x	2	x	5
12 × 12	Bar-Diagonal			x	10
16 × 12	Bar-Diagonal			x	12
24 × 12	Bar-Diagonal			x	15
ADJUSTABLE					
7 × 14	Screen	x	4		
6 × 8	Screen or plain	x	2		
8 × 12	Screen or plain	x	4	x	8
8 × 16	Screen or plain	x	5	x	11

Numerous other sizes available

VENTILATORS—WALL

Style	Size	Open Area (Sq. In.)	Weight (Lbs.)
Triangular	29 × 11	84	2
Triangular	Adjustable	75	2 1/2
Triangular	36 × 9	70	2
Square	8 × 8	35	3 1/2
Rectangular	8 × 16	91	4 1/2
Square	12 × 12	99	5
Rectangular	12 × 18	165	6 1/2
Rectangular	18 × 24	330	8 1/2
Rectangular	8 × 24	160	7 1/2
Quarter Circle	14 × 15	56	7 1/2
Half Circle	28 × 15	116	12 1/2

Use 50 sq. inches net free vent area per 100 sq. ft. floor area

VENTS—GAS

Sizes (In.)	Lengths (In.)	Wt. (Lbs./Ft.)
3	6-18-36-60	1/2
4	6-18-36-60	3/4
5	6-18-36-60	1
6	6-18-36-60	1 1/4
7	6-18-36	1 1/2
8	6-18-36	1 3/4

WALLBOARD—AREAS COVERED

No. of Pcs. 4' Wide (Sq. Ft.)	Lengths				
	6 Ft.	7 Ft.	8 Ft.	9 Ft.	10 Ft.
1	24	28	32	36	40
2	48	56	64	72	80
3	72	84	96	108	120
4	96	112	128	144	160
5	120	140	160	180	200
6	144	168	192	216	240
7	168	196	224	252	280
8	192	224	256	288	320
9	216	252	288	324	360
10	240	280	320	360	400
15	360	420	480	540	600
20	480	560	640	720	800
25	600	700	800	900	1000
30	720	840	960	1080	1200
35	840	980	1120	1260	1400
40	960	1120	1280	1440	1600
45	1080	1260	1440	1620	1800
50	1200	1400	1600	1800	2000
55	1320	1540	1760	1980	2200
60	1440	1680	1920	2160	2400
65	1560	1820	2080	2340	2600
70	1680	1960	2240	2520	2800
75	1800	2100	2400	2700	3000

WALLBOARD—GYPSUM
(JOINT COMPOUND REQUIREMENTS)

Square Feet	Prepared Joint Compound		Joint Tape (Rolls)	
	1 Gal.	5 Gal.	60'	250'
100 – 200	1		2	
300 – 400	2		3	
400 – 600	3			1
600 – 800	4		1	1
800 – 1000		1	2	1
			(Or (1) 500 ft. roll)	

Powder Joint Compound: 30 lbs. required/500 sq. ft.
Adhesive installation requires 1 quart sized tube/500 sq. ft.

WALLBOARD—GYPSUM (MAXIMUM FRAMING SPACING)

Single Ply Gypsum Board Thickness (In.)	Application to Framing (In.)	Maximum o.c. Spacing of Framing (In.)
CEILINGS		
†3/8	*Perpendicular	16
1/2	Perpendicular	16
1/2	*Parallel	16
5/8	Parallel	16
1/2	*Perpendicular	24
5/8	Perpendicular	24
SIDEWALLS		
3/8	Perpendicular or Parallel	16
1/2	Perpendicular or Parallel	24
5/8	Perpendicular or Parallel	24

†Should not support thermal insulation.

*On ceilings to receive a water base texture material, either hand or spray applied, install gypsum board perpendicular to framing and increase board thickness from 3/8 in. to 1/2 in. for 16 in. o.c. framing and from 1/2 in. to 5/8 in. for 24 in. o.c. framing.

WALLBOARD—GYPSUM (SIZES, SPECIFICATIONS)

Thickness (In.)	Edges	Width (Ft.)	Length (Ft.)	App. Weight 1000 Sq. Ft. (Lbs.)	Stud Spacing (In.)	Joint Treatment	Nail and Size	Nails 1000 Sq. Ft. (Lbs.)	Nail Spacing
1/2	Recessed	4	7, 8, 10 & 12 – 14	2100	16 or 24 o.c.	Tape Cement	5d Cement Coated	6	8" Walls 7" Ceilings
3/8	Recessed	4	7, 8, 10 & 12 – 14	1550	16 o.c.	Tape Cement	4d Cement Coated	4 1/2	8" Walls 7" Ceilings
3/8	Beveled	4	8, 10 & 12 – 14	1550	16 o.c.	No treatment required	4d Cement Coated	4 1/2	8" Walls 7" Ceilings
1/4	Square	4	7, 8, 9, & 10 – 12	1100	16 o.c.	Tape or Batterns or Panel Strips	4d Cement Coated 6d over existing plaster	4d-4 1/2 6d-6	8" Walls 7" Ceilings
1	Square	2	8, 9, 10 – 12	4000	—	For Laminated Partitions			

Data based on wood stud application. Check codes and local regulations on spacing, power-driven screw nails etc. for steel stud construction.

Some sizes available in foil back type

WALLBOARD—GYPSUM (TYPES, USES)

- Standard: Gray liner paper packing, smooth cream surface.
- Insulating: Aluminum foil bonded to backing.
- Predecorated: Decorated vinyl or paper bonded to surface.
- Type X: Fire-resistant core for high fire-resistant ratings.
- Backing: Gray liner both sides for use in multiple ply construction for fire resistance, sound control, and wall strength. Available in Type X, water resistant and insulating types.
- Edge types: Tapered, beveled, square, rounded.

Installed by: Nailing, power screws and adhesives, using joint compound and joint tape.

WALLBOARD—GYPSUM FRAME SPACING
For Ceiling Panels to be Covered with Water-Based Texturing Materials

Board Thickness	Application Method (Long Edge Relative to Frame)	Max. Framing Spacing o.c.
3/8	not recommended	—
1/2	perpendicular only	16
5/8	perpendicular only	24

Note: For adhesively laminated double-layer applications with 3/4″ or more total thickness, 24″ o.c. max.

Water-based texturing materials applied to ceilings should be completely dry before insulation and vapor retarder are installed.

WALLBOARD—GYPSUM FRAME SPACING (TWO-PLY APPLICATION WITHOUT ADHESIVE BETWEEN PLIES)

Gypsum Board Thickness (In.)		Application to Framing		Maximum o.c. Spacing of Framing (In.)
Base	Face	Base	Face	
CEILINGS				
3/8	3/8	Perpend	Perpend*	16
1/2	3/8	Parallel	Perpend*	16
1/2	1/2	Parallel	Perpend	16
1/2	1/2	Perpend	Perpend*	24
5/8	1/2	Perpend	Perpend*	24
5/8	5/8	Perpend	Perpend*	24
SIDEWALLS				

For two-ply application with no adhesive between plies, 3/8 in., 1/2 in., or 5/8 in. thick gypsum board may be applied perpendicular or parallel on framing spaced a maximum of 24 in. o.c. Maximum spacing should be 16 in. o.c. when 3/8 in. thick board is used as face ply.

*On ceilings to receive a water base texture material, either hand or spray applied, install gypsum board perpendicular to framing and increase board thickness from 3/8 in. to 1/2 in. for 16 in. o.c. framing and from 1/2 in. to 5/8 in. for 24 in. o.c. framing.

WALLBOARD—GYPSUM FRAME SPACING (TWO-PLY APPLICATION WITH ADHESIVE BETWEEN PLIES*)

Gypsum Board Thickness (In.)		Application to Framing		Maximum o.c. Spacing of Framing (In.)
Base	Face	Base†	Face†	
CEILINGS				
3/8	3/8	Perp	Perp or Par	16
1/2	3/8	Perp or Par	Perp or Par	16
1/2	1/2	Perp or Par	Perp or Par	16
5/8	1/2	Par	Perp or Par	24
5/8	5/8	Perp or Par	Perp or Par	24
SIDEWALLS				

For two-ply application with adhesive between plies, 3/8 in., 1/2 in. or 5/8 in. thick gypsum board may be applied perpendicular or parallel on framing spaced a maximum of 24 in. o.c.

*Adhesive between plies should be dried or cured prior to any decorative treatment. This is especially important when water based texture material (hand or spray applied) is to be used.
† Perp = perpendicular
  Par = parallel

WALLPAPER—REQUIREMENTS
Single Rolls Based on 36 Square Feet/Roll

Size of Room	Ceiling Heights			Yds. of Border	Rolls for Ceiling
	8 Ft.	9 Ft.	10 Ft.		
4 × 8	6	7	8	9	2
4 × 10	7	8	9	11	2
4 × 12	8	9	10	12	2
6 × 10	8	9	10	12	2
6 × 12	9	10	11	13	3
8 × 12	10	11	13	15	4
8 × 14	11	12	14	16	4
10 × 14	12	14	15	18	5
10 × 16	13	15	16	19	6
12 × 16	14	16	17	20	7
12 × 18	15	17	19	22	8
14 × 18	16	18	20	23	8
14 × 22	18	20	22	26	10
15 × 16	15	17	19	23	8
15 × 18	16	18	20	24	9
15 × 20	17	20	22	25	10
15 × 23	19	21	23	28	11
16 × 18	17	19	21	25	10
16 × 20	18	20	22	26	10
16 × 22	19	21	23	28	11
16 × 24	20	22	25	29	12
16 × 26	21	23	26	31	13
17 × 22	19	22	24	28	12
17 × 25	21	23	26	31	13
17 × 28	22	25	28	32	15
17 × 32	24	27	30	35	17
17 × 35	26	29	32	37	18
18 × 22	20	22	25	29	12
18 × 25	21	24	27	31	14
18 × 28	23	26	28	33	16
20 × 26	23	26	28	33	17
20 × 28	24	27	30	34	18
20 × 34	27	30	33	39	21

Deduct for openings-allow for matching

WALLS—FRAME

Sheathing Material	Relative Rigidity	Relative Strength
1 × 8″ Diagonal Sheathing	1.0	1.3
²⁹/₃₂″ Fiberboard. . . . . (8d nails, spaced 3″ at all vertical edges, 5 ¹/₂″ to 6″ elsewhere)	1.6	2.1
Horizontal Sheathing. . . . . (1 × 8 sheathing, 1 × 4 let in braces; 8d nails, 2 per stud crossing)	1.5	2.2
¹/₄″ Plywood Nailed . . . . (6d nails spaced 5″ at edges, 10″ elsewhere)	2.0	2.8
¹/₄″ Plywood . . . . Glued to frame	3.7	4.0

WALLS—PLASTER PARTITION
(MATERIALS REQUIRED 1100 SQUARE YARDS)
2-Inch Solid Wall Channels Lath Plaster 2 Sides

| Item | 12″ | Channel Spacing | |
		16″	19″
Metal Lath (Sq. Yds.)	100 – 105	100 – 105	100 – 105
Tie Wire (Pounds)	15	11 ¹/₄	9 ¹/₂
Channels: depend upon height			

WALLS—STEEL & PLASTER PARTITION

Total Thickness (In.)	Channel or Stud Size (In.)	Height Limits (Ft.)	Length Limits (Ft.)
2	³/₄	8	None
2	³/₄	12	None
2	³/₄	14	24
2 ¹/₄	³/₄	16	32
2 ¹/₂	³/₄	18	26
2 ³/₄	³/₄	20	30
2	1 ¹/₂	24	36
3 ¹/₂	1 ¹/₂	30	30

Check manufacturer's specifications

WALLS—STEEL STUD SPACING

Spacing (In.)	Lath Types
12	2.5 lb. Diamond Mesh Lath
16	3.4 lb. Diamond Mesh Lath
16	2.75 lb. $1/8''$ Riblath
16	$3/8''$ Plain or Perforated
19	3.4 lb. $1/8''$ Riblath
24	3.4 lb. $3/8''$ Riblath
24	4.0 lb. $3/8''$ Riblath

Check local codes

WALLS—STEEL STUDS & TRACK

Size (In.)	Ga.	Lengths (Ft.)	Wt./1000 Ft.
2 $1/2$	16	(8 – 9)	400
3 $1/4$	16	(10 – 12)	570
4	16	(14 – 16)	640
6	16	(18 – 20)	870

Widths Track 2 $1/2''$ × 3 $1/4''$ - 4''-6''
Lengths 8'-2''

Numerous types available

Walls—Weights/Square Foot, see page 310.

WASHERS—PLATE

Square Size		Bolt Size	Wt. (Lbs.)/100
$1/8''$	2'' × 2''	$1/2$	15
$3/16''$	3'' × 3''	$3/4$	40
$1/4''$	3'' × 3''	1	50

WALLS—WEIGHTS/SQUARE FOOT

Construction	Wt. (Lbs./ Sq. Ft.)
2 × 4″ studs, ¹/₂″ Building Board on both sides - unplastered	4
2 × 4″ studs, ¹/₂″ Insulating Lath on both sides, plastered each side ¹/₂″ plaster	13
2 × 4″ staggered studs, ¹/₂″ Insulating Lath both sides, plastered each side ¹/₂″ plaster	13
2 × 2″ special double stud construction, ¹/₂″ Insulating Lath and ¹/₂″ plaster both sides, ¹/₂″ Building Board standing loose in middle	14
Double 3″ gypsum hollow tile, plastered, ¹/₂″ Building Board standing free in ¹/₂″ air space separation	32
2 × 4″ studs, wood lath and plaster	17
1 ¹/₂″ solid metal lath and plaster	14
2″ solid gypsum tile, plastered both sides	20
4″ hollow clay tile, plastered both sides	27
8″ brick wall, ¹/₂″ plaster both sides, no furring	88
3″ Cinder Block, ⁵/₈″ plaster on both sides	32
4″ Cinder Block, ⁵/₈″ plaster on both sides	36
4″ Cinder Block 1″ plaster	32
8″ Expanded slag, 1″ plaster	56
4″ Substitute lightweight concrete block, ¹/₂″ plaster on both sides	30
8″ Substitute lightweight concrete block, unplastered	28
8″ Substitute lightweight concrete block, ¹/₂″ plaster on both sides	40
Cavity wall, two 4″ substitute lightweight concrete block, ¹/₂″ plaster on one inner face	45
4″ Pumice block, ¹/₂″ plaster on both sides	25
4″ Pumice block, ¹/₂″ plaster on one side only	20
4″ Substitute lightweight concrete block, ¹/₂″ plaster on both sides	31
8″ Substitute lightweight concrete block, ¹/₂″ plaster on both sides	47
3″ Substitute lightweight concrete block, painted with 2 coats cement paint	17
4″ Substitute lightweight concrete block, unpainted	17
4″ Substitute lightweight concrete block, painted with 2 coats cement paint	17
6″ Substitute lightweight concrete block, unpainted	21
6″ Substitute lightweight concrete block, painted with 2 coats cement paint	21
Cavity wall, two 3″ substitute lightweight concrete block, ³/₈″ plaster on one unexposed face	17

WATERPROOFING—CLEAR LIQUID

Coverage	1 gal. cans
	5 gal. cans
	30 – 55 gal. drums
Brickwork	100 – 200 ft./gal.
Concrete	200 – 300 ft./gal.
Concrete Block	100 – 200 ft./gal.

WATERPROOFING—MEMBRANE
Plies Required for Various Water Pressures

Head of Water (Ft.)	Felt, Fabric, or Combined Felt and Fabric		Lbs.
	Plies Required	Pitch Mopping Required	
1 – 3	2	3	105
3 – 6	3	4	140
6 – 9	4	5	175
9 – 12	5	6	210
12 – 18	6	7	245
18 – 25	7	8	280
25 – 35	10	11	315
35 – 50	11	12	350
50 – 75	13	14	420
75 – 100	14	15	455

WEIGHT & SPECIFIC GRAVITY—LIQUIDS

Liquids at 32°F.	Weight of One Cu. Ft. Lbs.	Weight of One Gal. (Imperial) Lbs.	Specific Gravity Water - 1
Mercury.................	848.7	136.0	13.596
Bromine	185.1	29.7	2.966
Sulphuric acid............	114.9	18.4	1.84
Nitrous acid..............	96.8	15.5	1.55
Chloroform	95.5	15.3	1.53
Water of the Dead Sea	77.4	12.4	1.24
Nitric acid	76.2	12.2	1.22
Acetic acid	67.4	10.8	1.08
Milk.....................	64.3	10.3	1.03
Sea water	64.05	10.3	1.026
Pure water (distilled) at 39°F	62.425	10.0	1.0
Oil, linseed	58.7	9.4	0.94
Oil, poppy	58.1	9.3	0.93
Oil, rape seed............	57.4	9.2	0.92
Oil, whale	57.4	9.2	0.92
Oil, olive	57.1	9.15	0.915
Oil, turpentine...........	54.3	8.7	0.87
Oil, potato	51.2	8.2	0.82
Petroleum	54.9	8.8	0.88
Naphtha	53.1	8.5	0.85
Ether, nitric	69.3	11.1	1.11
Ether, sulphurous	67.4	10.8	1.08
Ether, nitrous.............	55.6	8.9	0.89
Ether, acetic	55.6	8.9	0.89
Ether, hydrochloric	54.3	8.7	0.87
Ether, sulphuric...........	44.9	7.2	0.72
Alcohol, proof spirit........	57.4	9.2	0.92
Alcohol, pure.............	49.3	7.9	0.79
Benzine.................	53.1	8.5	0.85
Wood Spirit	49.9	8.0	0.80

WEIGHT & SPECIFIC GRAVITY—METALS

Metal	Specific Gravity Range According to Several Authorities	Specific Gravity Approx. Mean Value Used in Calculation of Weight	Weight/ Cu. Ft. Lbs.	Weight/ Cu. In. Lbs.
Aluminum	2.56 – 2.71	2.67	166.5	.0963
Antimony	6.66 – 6.86	6.76	421.6	.2439
Bismuth	9.74 – 9.90	9.82	612.4	.3544
Brass, copper and zinc)				
80 20)		8.60	536.3	.3103
70 30)		8.40	523.8	.3031
60 40)	7.8 – 8.6	8.36	521.3	.3017
50 50)		8.20	511.4	.2959
Bronze (copper 95 to 80)	8.52 – 8.96	8.853	552.0	.3195
(tin 5 to 20)				
Cadmium	8.6 – 8.7	8.65	539.0	.3121
Calcium	1.58			
Chromium	5.0			
Cobalt	8.5 – 8.6			
Gold, pure	19.245 – 19.361	19.258	1200.9	.6949
Copper	8.69 – 8.92	8.853	552.0	.3195
Iridium	22.38 – 23.0		1396.0	.8076
Iron, cast	6.85 – 7.48	7.218	450.0	.2604
Iron, wrought	7.4 – 7.9	7.70	480.0	.2779
Lead	11.07 – 11.44	11.38	709.7	.4106
Manganese	7.0 – 8.0	8.00	499.0	.2887
Magnesium	1.69 – 1.75	1.75	109.0	.0641
Mercury (32°)	13.60 – 13.62	13.62	849.3	.4915
Mercury (60°)	13.58	13.58	846.8	.4900
(212°)		13.38	834.4	.4828
Nickel	8.279 – 8.93	8.8	548.7	.3175
Platinum	20.33 – 22.07	21.5	1347.0	.7758
Potassium	0.865			
Silver	10.474 – 10.511	10.505	655.1	.3791
Sodium	0.97			
Steel	7.69 – 7.932	7.854	489.6	.2834
Tin	7.29 – 7.409	7.350	458.3	.2662
Titanium	5.3			
Tungsten	17.0 – 17.6			
Zinc	6.86 – 7.20	7.00	436.5	.2526

WEIGHT & SPECIFIC GRAVITY—NONMETALLIC

Substance	Specific Gravity	Av. Wt. Lb./ Cu. Ft.
Asbestos	2.1 – 2.8	153
Ashes	...	43
Asphaltum	1.39	87
Barytes	4.50	281
Basalt	2.7 – 3.2	184
Bauxite	2.55	159
Bluestone	2.2 – 2.5	147
Brick, soft	1.6	100
Brick, common	1.79	112
Brick, hard	2.0	125
Brick, pressed	2.16	135
Brick, fire	2.24 – 2.4	145
Brick, sand-lime	2.18	136
Brickwork, mortar	1.6	100
Brickwork, cement	1.79	112
Borax	1.7 – 1.8	112
Cement, portland, loose	1.44	92
Chalk	1.8 – 2.6	137
Clay	1.92 – 2.4	137
Coal, anthracite	1.4 – 1.8	97
Coal, bituminous	1.2 – 1.5	84
Coal, lignite	1.1 – 1.4	78
Coal, charcoal	0.27 – 0.58	18
Coke	1.0 – 1.4	22 – 27
Concrete	1.92 – 2.48	133
Cork	0.22 – 0.26	15
Dolomite	2.9	181
Earth, dry, loose	1.2	75
Earth, dry, packed	1.5	93
Earth, moist, loose	1.3	81
Earth, moist, packed	1.6	100
Earth, mud, flowing	1.7	106
Earth, mud, packed	1.8	112
Emery	4.0	250
Feldspar	2.5 – 2.6	159
Glass, common	2.5 – 2.75	164
Glass, crystal	2.90 – 3.00	184
Glass, flint	2.9 – 2.31	188
Glass, plate	2.45 – 2.72	161
Gneiss	2.4 – 2.7	165
Granite	2.5 – 3.1	179
Graphite	1.9 – 2.3	126
Gravel, dry, loose	1.4 – 1.7	90 – 105
Gravel, dry, packed	1.6 – 1.9	100 – 120
Gravel, wet	1.9	120
Greenstone	2.8 – 3.2	187
Gypsum	2.08 – 2.4	140
Hornblende	3.2 – 3.52	210
Ice	0.88 – 0.92	55-57
Leather	0.86 – 1.02	59

Substance	Specific Gravity	Av. Wt. Lb./ Cu. Ft.
Lime, quick, in bulk	0.8 – 0.96	55
Limestone	2.3 – 2.9	100
Magnesia, carbonate	2.4	150
Magnesite	3.0	187
Marble	2.56 – 2.88	170
Masonry, dry, rubble	2.24 – 2.56	150
Masonry, dressed	2.24 – 2.88	160
Mica	2.80	175
Mortar	1.44 – 1.6	95
Mud	1.7 – 1.8	111
Paper	0.70 – 1.15	58
Paraffin	0.87 – 0.91	56
Peat	0.65 – 0.85	47
Phosphate rock	3.2	200
Pitch	1.15	72
Plaster-of-Paris	1.5 – 1.8	103
Porcelain	2.3 – 2.5	250
Porphyry	2.6 – 2.9	172
Pumice	0.37 – 0.90	40
Riprap, limestone	1.3 – 1.4	80-85
Riprap, sandstone	1.4	90
Riprap, shale	1.7	105
Rubber, caoutchouc	0.92 – 0.96	59
Rubber, manufactured	1.0 – 2.0	95
Salt	1.12	80
Saltpeter	1.07	67
Sand	1.44 – 1.76	100
Sand, wet	1.89 – 2.07	125
Sandstone	2.24 – 2.4	145
Serpentine	2.4 – 2.7	165
Shale	2.6 – 2.9	172
Slag, bank	1.1 – 1.2	69
Slag, bank screenings	1.5 – 1.9	107
Slag, machine	1.5	96
Slag, sand	0.8 – 0.9	53
Slate	2.72 – 2.88	175
Soapstone	2.65 – 2.8	170
Starch	1.53	96
Stone, various	2.16 – 3.4	135-200
Stone, crushed	1.6	100
Sulphur	1.93 – 2.07	125
Talc	2.6 – 2.8	169
Tar, bituminous	1.20	75
Terra-cotta	1.9	119
Tile	1.76 – 1.92	115
Trap rock	2.72 – 3.4	185
Wool	1.32	82

WEIGHTS—BUILDING MATERIALS

Material	Lb./Cu. Ft.
Asphalt-pavement composition	100
Birch	48
Bluestone	160
Brick, best pressed	150
Brick, common and hard	125
Brickwork in lime mortar, av.	120
Brickwork in cement mortar, av.	130
Brickwork, pressed brick, thin joints	140
Clay, dry	63 to 95
Clay, fire	130
Clay, wet	120 to 140
Clay, tile	60
Concrete -	
Gravel and stone aggregate	150
Lightweight aggregate	100
Aerated	50 to 90
Concrete masonry -	
Gravel and stone aggregate	80
Lightweight aggregate	50
Crushed stone	35 to 90
Cypress	36
Fill dirt	90 to 110
Fir, douglas	36
Firebrick	150
Granite	167
Gravel	95 to 100
Gypsum partition block (hollow)	48
Hemlock	30
Hollow tile partition block	60
Insulation - loose fill	10
Iron, cast	450
Iron, wrought	480
Limestone	155 to 172
Maple	48
Marble	171 to 179
Masonry, squared granite or limestone	165
Masonry, granite or limestone, dry rubble	138
Masonry, granite or limestone, rubble	150
Masonry, sandstone	150
Mineral wool	12
Mortar, hardened	90 to 100
Mortar	120
Oak	48
Perlite concrete	45 to 60
Plaster	96
Sand, dry	100
Slate	172 to 177
Spruce	30
Steel, structural	489.6
Steel	490
Stone masonry	160

Material	Lb./Cu. Ft.
Stone riprap	65
Terra cotta, solid	120
Terra cotta, masonry work	70 to 80
Tile, solid	110 to 120
Topsoil	80 to 90
Vermiculite concrete	45 to 60
Wood, dry, yellow pine	48

WIRE—BLACK, GALVANIZED (SIZES, WEIGHTS)

	Approx. Ft./ 100 Lbs.	Approx. Weight/Ft.
1	470	.213
2	532	.188
3	600	.167
4	741	.135
5	877	.114
6	1020	.098
7	1200	.084
8	1430	.070
9	1700	.059
10	2060	.049
11	2580	.039
12	3370	.030
13	4450	.022
14	5860	.017
15	7220	.014
16	9600	.010
17	12800	.008
18	16600	.006
19	22300	.0045
20	31000	.0032

Coils 6 to 20 ga. 16" - 22" I.S. dia wts 80 to 150#

WIRE—HANGER

Ga.	Dia.	Wt./100 Ft.	Packed
8	.165″	7 1/2 lbs.	50 – 100 lb. coils

WIRE MESH—REINFORCEMENT (CONCRETE)

Common Sizes	Ga.	Width (In.)	Length (Ft.)	Sq. Ft./ Roll	Wt./ 100 Sq. Ft.
6 × 6	10	60	150	750	21 lbs.
6 × 6	8	60	150	750	30 lbs.
6 × 6	6	60	150	750	42 lbs.

Sizes available range from 2″ × 2″ mesh, 16 ga. wire to 6″ × 12″ mesh, 3 ga.

WIRE MESH—REINFORCEMENT (CONCRETE MASONRY)
Small Mesh Galvanized Wire 200 Linear Feet/Roll

Wall Size (In.)	Wt. (Lbs)/1000 Lin. Ft.
4	83
6	98
8	113
10	135
12	157

WIRE MESH—REINFORCEMENT (STUCCO)

Type	Opening Size (In.)	Gauge	Weight (Lbs./Sq. Ft.)
Woven Wire	1 Hex	18	1.7
	1 1/2 Hex	17	1.4
Welded Wire	2 × 2	16	1.4
	1 × 1	18	1.4
	1 1/2 × 1 1/2	16	1.9

(3 × 150 ft. rolls)

WIRE MESH—SIZES, WEIGHTS

Mesh Wire Size	Wt. (Lbs.)/ 100 Sq. Ft.	Mesh Wire Size	Wt. (Lbs.)/ 100 Sq. Ft.
2 × 2 $^{16}/_{16}$	13	4 × 8 $^{12}/_{14}$	12
2 × 2 $^{14}/_{14}$	21	4 × 8 $^{12}/_{12}$	14
2 × 2 $^{13}/_{13}$	28	4 × 8 $^{11}/_{12}$	17
2 × 2 $^{12}/_{12}$	37	4 × 8 $^{10}/_{12}$	20
2 × 2 $^{11}/_{11}$	48	4 × 8 $^{9}/_{12}$	23
2 × 2 $^{10}/_{10}$	60	4 × 8 $^{8}/_{12}$	27
2 × 4 $^{14}/_{14}$	16	4 × 8 $^{7}/_{11}$	33
2 × 4 $^{13}/_{14}$	19	4 × 12 $^{11}/_{12}$	16
2 × 4 $^{12}/_{12}$	28	4 × 12 $^{12}/_{12}$	13
2 × 4 $^{11}/_{11}$	36	4 × 12 $^{10}/_{12}$	19
2 × 12 $^{3}/_{8}$	105	4 × 12 $^{9}/_{12}$	22
2 × 12 $^{0}/_{6}$	166	4 × 12 $^{8}/_{12}$	25
2 × 16 $^{8}/_{12}$	46	4 × 12 $^{7}/_{11}$	31
2 × 16 $^{7}/_{11}$	55	4 × 12 $^{6}/_{10}$	36
2 × 16 $^{6}/_{10}$	65	4 × 12 $^{5}/_{10}$	42
2 × 16 $^{5}/_{10}$	75	4 × 12 $^{5}/_{7}$	45
2 × 16 $^{4}/_{9}$	89	4 × 12 $^{4}/_{9}$	49
2 × 16 $^{3}/_{8}$	104	4 × 16 $^{10}/_{12}$	18
2 × 16 $^{2}/_{8}$	119	4 × 16 $^{9}/_{12}$	21
2 × 16 $^{1}/_{7}$	139	4 × 16 $^{8}/_{12}$	25
3 × 3 $^{10}/_{10}$	41	4 × 16 $^{7}/_{11}$	30
3 × 3 $^{9}/_{9}$	49	4 × 16 $^{6}/_{10}$	35
3 × 3 $^{8}/_{8}$	58	4 × 16 $^{5}/_{10}$	40
3 × 16 $^{8}/_{12}$	32	4 × 16 $^{4}/_{9}$	48
3 × 16 $^{7}/_{11}$	38	4 × 16 $^{3}/_{8}$	56
3 × 16 $^{6}/_{10}$	45	4 × 16 $^{2}/_{8}$	64
3 × 16 $^{5}/_{10}$	52	6 × 6 $^{10}/_{10}$	21
3 × 16 $^{4}/_{9}$	61	6 × 6 $^{9}/_{9}$	25
3 × 16 $^{3}/_{8}$	72	6 × 6 $^{8}/_{8}$	30
3 × 16 $^{2}/_{8}$	83	6 × 6 $^{7}/_{7}$	36
3 × 16 $^{1}/_{7}$	96	6 × 6 $^{6}/_{6}$	42
3 × 16 $^{0}/_{6}$	113	6 × 6 $^{5}/_{5}$	49
4 × 4 $^{14}/_{14}$	11	6 × 6 $^{4}/_{6}$	50
4 × 4 $^{13}/_{13}$	14	6 × 6 $^{4}/_{4}$	58
4 × 4 $^{12}/_{12}$	13	6 × 6 $^{3}/_{3}$	68
4 × 4 $^{10}/_{10}$	31	6 × 6 $^{2}/_{2}$	78
4 × 4 $^{8}/_{8}$	44	6 × 6 $^{1}/_{1}$	91
4 × 4 $^{7}/_{7}$	53	6 × 6 $^{0}/_{0}$	107
4 × 4 $^{6}/_{6}$	62	6 × 12 $^{7}/_{7}$	27
4 × 4 $^{4}/_{4}$	85	6 × 12 $^{6}/_{6}$	32
4 × 8 $^{13}/_{13}$	11	6 × 12 $^{5}/_{5}$	37

WIRE ROPE—PRESTRESSED CONCRETE (PROPERTIES)

Strand Diam. (In.)	Wt./ 1000' (Lbs.)	Nominal Area (Sq. In.)	Ultimate Strength (Lbs.)	Yield Strength (Lbs.)
1/4	122	.036	9,000	7,650
5/16	198	.058	14,500	12,300
3/8	274	.080	20,000	17,000
7/16	373	.109	27,000	23,000
1/2	494	.144	36,000	30,600

WIRE ROPE—STRESS-RELIEVED PRESTRESSED CONCRETE (PROPERTIES)

Size (Dia. In.)	Area (Sq. In.)	Weight (Lbs./100 Ft.)	Minimum Ultimate Strength		Minimum Yield Strength	
			(PSI)	(Lbs.)	(PSI)	(Lbs.)
0.276	.0598	203.2	235,000	14,050	188,000	11,240
0.250	.0491	166.7	240,000	11,780	192,000	9,430
0.196	.0302	102.5	250,000	7,550	200,000	6,040
0.192	.0289	98.32	250,000	7,220	200,000	5,780

WIRE—TIE

Gauge	Sizes	Packed	Ft./ 100 Lbs.
18	36" Lengths	25 lb. Hanks	16,600
18	Coil	50 – 1000 lbs.	16,600
16	Coil	50 – 1000 lbs.	9,600

Packaging will vary

WOOD—PROPERTIES

Type Wood	Wt./Cu. Ft.		Based on Redwood 100%				
	Green (Lbs.)	Dry (Lbs.)	Bonding Str. %	Comp. Str. %	Stiffness %	Hardness %	Shock Resistance %
Baldcypress (Taxodium distichum)	50	32	95	89	99	96	117
Cedar, northern white (Thuja occidentalis)	28	22	60	50	57	56	72
Douglas-fir (Pseudotsuga glauca)	35	30	90	81	104	96	103
Douglas-fir (Pseudotsuga menziesii)	38	34	108	104	132	109	125
Fir, grand (Abies grandis)	44	28	87	80	114	80	111
Fir, noble (Abies procera)	30	26	89	74	109	72	105
Fir, pacific (Abies amabilis)	36	27	84	74	107	69	108
Fir, white (Abies concolor)	47	26	87	71	93	78	92
Firs, white (Average of 4 species)	41	26	87	74	103	76	102
Hemlock, western (Tsuga heterophylla)	41	29	89	82	105	93	112
Pine, loblolly (Pinus taeda)	54	38	112	101	121	115	143
Pine, longleaf (Pinus palustris)	50	41	128	119	138	141	158
Pine, northern white (Pinus strobus)	36	25	76	65	87	65	85
Pine, ponderosa (Pinus ponderosa)	45	28	78	67	82	76	89
Pine, shortleaf (Pinus echinata)	51	38	117	101	124	126	171
Pine, sugar (Pinus lambertiana)	51	25	77	66	82	70	85
Pine, western white (Pinus monticola)	35	27	83	73	100	65	100
Port Orford-cedar (Chamaecyparis lawsoniana)	36	29	99	87	123	89	122
Red cedar, eastern (Juniperus virginiana)	37	33	81	84	58	150	175
Red cedar, western (Thuja plicata)	27	23	72	72	79	70	80
Redwood (Sequoia sempervirens)	52	28	100	100	100	100	100
Spruce, Sitka (Picea sitchensis)	33	28	87	73	105	81	117

WOOD—PROPERTIES

Types	Insulation Values	Stiffness	Nail Holding Value
	Resistance to Loss of Heat B.T.U./ Sq. Ft./Hour/1° Difference in Temp.	Lbs./Sq. Ft. Deflection at 1/360th at 24″ Space	Resistance to Nail Withdrawal. Load in Pounds Full Penetration
Softwoods 3/4″	.98	212	75
Plywoods 1/2″	.60	66	60
Plywoods 3/8″	.37	32	45
Fiberboards 1/2″	1.27	4	5
Fiberboards 25/52″	2.05	10	10

WOOD—RECOMMENDED MOISTURE CONTENT
Values for Various Wood Items at Time of Installation

Use of Wood	Moisture Content for—					
	Most Areas of United States		Dry Southwestern Area[1]		Damp, Warm, Coastal Areas[1]	
	Average[2]	Individual Pieces	Average[2]	Individual Pieces	Average[2]	Individual Pieces
	Pct.	Pct.	Pct.	Pct.	Pct.	Pct.
Interior: Woodwork, flooring, furniture wood trim, laminated timbers, cold-press plywood	8	6 – 10	6	4 – 9	11	8 – 13
Exterior: Siding, wood trim, framing, sheathing, laminated timbers	12	9 – 14	9	7 – 12	12	9 – 14

[1]Major areas are indicated on on page 323.
[2]To obtain a realistic average, test at least 10 pct. of each item. If the amount of a given item is small, several tests should be made. For example, in an ordinary dwelling having about 60 floor joists, at least 10 tests should be made on joists selected at random.

WOOD—RECOMMENDED MOISTURE CONTENT

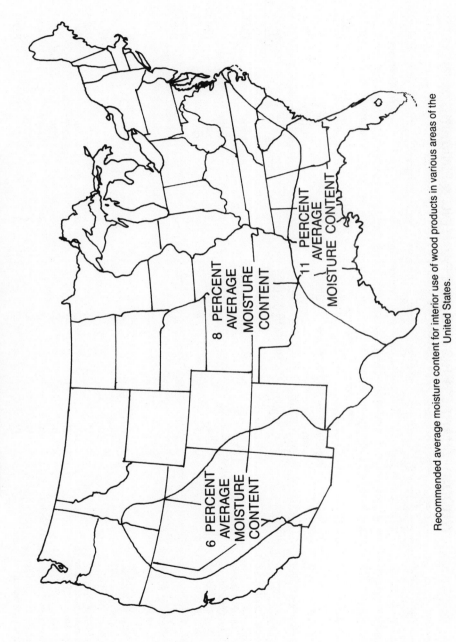

8 PERCENT AVERAGE MOISTURE CONTENT

11 PERCENT AVERAGE MOISTURE CONTENT

6 PERCENT AVERAGE MOISTURE CONTENT

Recommended average moisture content for interior use of wood products in various areas of the United States.

Woods—Average Comparative Properties, see page 326.

WOODS—CHARACTERISTICS FOR PAINTING & FINISHING

Omissions in the Table Indicate Inadequate Data for Classification

Wood	Ease of Keeping Well Painted; (Scale: I—Easiest, V—Most Exacting)[1]	Weathering		Appearance	
		Resistance to Cupping; (Scale: 1—Best, 4—Worst)	Conspicuous of Checking; (Scale: 1—Least, 2—Most)	Color of Heartwood (Sapwood Is Always Light)	Degree of Figure on Flat-grained Surface
SOFTWOODS					
Cedar:					
Alaska-	I	1	1	Yellow	Faint
California incense-	I	—	—	Brown	"
Port-Orford-	I	—	1	Cream	"
Western redcedar-	I	1	1	Brown	Distinct
White-	I	1	—	Light brown	"
Cypress-	I	1	1	"	Strong
Redwood-	I	1	1	Dark brown	Distinct
Products[2] overlaid with resin-treated paper-	I	—	1	—	—
Pine:					
Eastern white-	II	2	2	Cream	Faint
Sugar-	II	2	2	"	"
Western white-	II	2	2	"	"
Ponderosa-	III	2	2	"	Distinct
Fir, commercial white-	III	2	2	White	Faint
Hemlock-	III	2	2	Pale brown	"
Spruce-	III	2	2	White	"
Douglas-fir (lumber and plywood)-	IV	2	2	Pale red	Strong
Larch-	IV	2	2	Brown	"
Lauan (plywood)-	IV	2	2	"	Faint

Kind of wood	Group [1]			Color	Figure
Pine:					
Norway	IV	2	2	Light brown	Distinct
Southern (lumber and plywood)	IV	2	2	"	Strong
Tamarack	IV	2	2	Brown	"

HARDWOODS

Kind of wood	Group [1]			Color	Figure
Alder	III	—	—	Pale brown	Faint
Aspen	III	2	1	"	"
Basswood	III	2	2	Cream	"
Cottonwood	III	4	2	White	"
Magnolia	III	2	1	Pale brown	"
Yellow-poplar	IV	4	2	"	"
Beech	IV	4	2	Light brown	"
Birch	IV	4	2	Brown	"
Cherry	IV	4	2	Brown	"
Gum	IV	4	2	Pale brown	"
Maple	IV	4	2	Light brown	"
Sycamore	V or III	—	—	"	Distinct
Ash	V or III	4	2	Light brown	Faint
Butternut	V or III	3	2	"	Distinct
Chestnut	V or III	3	2	Light brown	"
Walnut	V or IV	4	2	Dark brown	"
Elm	V or IV	4	2	Brown	"
Hickory	V or IV	4	2	Light brown	"
Oak, white	V or IV	4	2	Brown	"
Oak, red	V or IV	4	2	"	"

[1] Woods ranked in group V for ease of *keeping well painted* are hardwoods with large pores that need filling with wood filler for durable painting. When so filled before painting, the second classification recorded in the table applies.

[2] Plywood, lumber, and fiberboard with overlay or low-density surface.

WOODS—AVERAGE COMPARATIVE PROPERTIES OF CLEAR WOOD OF VIRGIN-GROWTH REDWOOD COMPARED WITH A NUMBER OF OTHER SPECIES

Commercial and Botanical Name of Species	Trees Tested	Specific Gravity, Oven-Dry, Based on Volume when Green	Weight/Cubic Foot		Shrinkage from Green to Oven-Dry Condition Based on Dimensions when Green		
			Green	At 12 Percent Moisture Content	Radial	Tangential	Volumetric (Composite Value)
1	2	3	4	5	6	7	8
	Number		Pounds	Pounds	Percent	Percent	Percent
Redwood (Sequoia sempervirens)	16	0.39	52	28	100	100	100
Cedar, Port Orford (Chamaecyparis lawsoniana)	14	.40	36	29	192	172	158
Cedar, eastern red (Juniperus virginiana)	5	.44	37	33	129	118	116
Cedar, western red (Thuja plicata)	15	.31	27	23	100	125	110
Cedar, northern white (Thuja occidentalis)	5	.29	28	22	88	118	103
Cypress, southern (Taxiodium distichum)	26	.42	50	32	158	155	155
Douglas fir (Pseudotsuga menziesii) (Coast type)	34	.45	38	34	208	195	181
Douglas fir (Pseudotsuga menziesii) (Rocky Mountain type)	10	.40	35	30	150	155	154
Fir, lowland white (Abies grandis)	10	.37	44	28	133	180	157
Fir, noble (Abies nobilis)	9	.35	30	26	188	208	188
Fir, silver (Abies amabilis)	6	.35	36	27	188	250	212
Fir, white (Abies concolor)	20	.35	47	26	133	175	142
Firs, white (Average of four species)	45	.35	41	26	158	198	164
Hemlock, western (Tsuga heterophylla)	18	.38	41	29	179	198	179
Pine, loblolly (Pinus taeda)	10	.50	54	38	229	188	190
Pine, longleaf (Pinus palustris)	34	.55	50	41	221	188	185
Pine, northern white (Pinus strobus)	18	.34	36	25	96	150	124
Pine, shortleaf (Pinus echinata)	12	.49	51	38	212	205	191
Pine, sugar (Pinus lambertiana)	9	.35	51	25	121	140	118
Pine, western white (Pinus monticola)	14	.36	35	27	171	185	176
Pine, ponderosa (Pinus ponderosa)	31	.38	45	28	162	158	145
Spruce, Sitka (Picea sitchensis)	25	.37	33	28	179	188	173

WOODS—DURABILITY COMPARISON OF WOODS COMMONLY USED FOR CONSTRUCTION

Wood Type	Durability of Heartwood		Weather Resistance		Volumetric Shrinkage to 10% Moisture Content
	Decay Resistance	Termite Resistance	Tendency to Weather Check	Tendency to Cup and Pull Nails Loose	
Redwood	High	*Resistant	Inconspicuous	Slight	4.5
Baldcypress	High	*Resistant	Inconspicuous	Slight	7.0
Western Redcedar	High	Nonresistant	Inconspicuous	Slight	4.5
Douglas-fir	Moderate to Low	Nonresistant	Conspicuous	Distinct	7.9
Pine (Western White)	Moderate to Low	Nonresistant	Conspicuous	Distinct	7.9
(Eastern White)	Moderate to Low	Nonresistant	Conspicuous	Distinct	5.5
(Ponderosa)	Moderate to Low	Nonresistant	Conspicuous	Distinct	6.4
(Southern Yellow)	Moderate to Low	Resistant (very resinous heartwood only)	Conspicuous	Distinct	8.2
(Sugar)	Moderate to Low	Nonresistant	Conspicuous	Distinct	5.3
Sitka Spruce	Moderate to Low	Nonresistant	Conspicuous	Distinct	7.7
Western Hemlock	Moderate to Low	Nonresistant	Conspicuous	Distinct	8.0

*Redwood and baldcypress are the only two woods available in commercial quantities that are naturally termite resistant. Comparisons cited above are based on comparisons in Wood Handbook, U.S.D.A. Forest Products Laboratory.

Woods—Grouping of, see page 330.

WOODS—MACHINING AND RELATED PROPERTIES OF SELECTED DOMESTIC HARDWOODS

Kind of Wood[1]	Planing— Perfect Pieces	Shaping— Good to Excellent Pieces	Turning— Fair to Excellent Pieces	Boring— Good to Excellent Pieces	Mortising— Fair to Excellent Pieces	Sanding— Good to Excellent Pieces	Steam Bending— Unbroken Pieces	Nail Splitting— Pieces Free from Complete Splits	Screw Splitting— Pieces Free from Complete Splits
	Pct.	Pct.	Pct.	Pct.	Pct.	Pct.	Pct.	Pct.	Pct.
Alder, red	61	20	88	64	52	.....	.....	.....	.....
Ash	75	55	79	94	58	75	67	65	71
Aspen	26	7	65	78	60	.....	.....	.....	.....
Basswood	64	10	68	76	51	17	2	79	68
Beech	83	24	90	99	92	49	75	42	58
Birch	63	57	80	97	97	34	72	32	48
Birch, paper	47	22	.....						
Cherry, black	80	80	88	100	100	64	56	66	60
Chestnut	74	28	87	91	70	19	44	82	78
Cottonwood	21	3	70	70	52	66	74	80	74
Elm, soft	33	13	65	94	75	.....	94	63	63
Hackberry	74	10	77	99	72	80	76	35	63
Hickory	76	20	84	100	98	37	85	73	76
Magnolia	65	27	79	71	32				

Species									
Maple, bigleaf	52	56	80	100	80	⋮	57	27	52
Maple, hard	54	72	82	99	95	38	59	58	61
Maple, soft	41	25	76	80	34	37	86	66	78
Oak, red	91	28	84	99	95	81	91	69	74
Oak, white	87	35	85	95	99	83	78	47	69
Pecan	88	40	89	100	98	⋮	67	69	69
Sweetgum	51	28	86	92	58	23	29	79	74
Sycamore	22	12	85	98	96	21			
Tanoak	80	39	81	100	100				
Tupelo, water	55	52	79	62	33	34	46	64	63
Tupelo, black	48	32	75	82	24	21	42	65	63
Walnut, black	62	34	91	100	98	⋯	78	50	59
Willow	52	5	58	71	24	24	73	89	62
Yellow-poplar	70	13	81	87	63	19	58	77	67

[1]Commercial lumber nomenclature.

WOODS—GROUPING OF SOME DOMESTIC WOODS ACCORDING TO HEARTWOOD DECAY

Resistant or Very Resistant	Moderately Resistant	Slightly or Nonresistant
Baldcypress (old growth)[1]	Bald cypress (young growth)[1]	Alder
Catalpa	Douglas-fir	Ashes
Cedars	Honey locust	Aspens
Cherry, black	Larch, western	Basswood
Chestnut	Oak, swamp chestnut	Beech
Cypress, Arizona	Pine, eastern white[1]	Birches
Junipers	Southern pine:	Buckeye
Locust, black[2]	Longleaf[1]	Butternut
Mesquite	Slash[1]	Cottonwood
Mulberry, red[2]	Tamarack	Elms
Oak:		Hackberry
Bur		Hemlocks
Chestnut		Hickories
Gambel		Magnolia
Oregon white		Maples
Post		Oak (red and black species)
White		Pines (other than longleaf, slash, and eastern white)
Osage orange[2]		Poplars
Redwood		Spruces
Sassafras		Sweetgum
Walnut, black		True firs (western and eastern)
Yew, Pacific[2]		Willows
		Yellow-poplar

[1]The southern and eastern pines and baldcypress are now largely second growth with a large proportion of sapwood. Consequently, substantial quantities of heartwood lumber of these species are not available.

[2]These woods have exceptionally high decay resistance.

Woods—Species and Typical Uses, see page 332.

WOODS—TYPICAL PROPERTIES

Kind of Wood	Working and Behavior Characteristics							Strength Properties			
	Hardness	Freedom from Warping	Ease of Working	Paint Holding	Nail Holding	Decay Resistance of Heartwood	Proportion of Heartwood	Bending Strength	Stiffness	Strength as a Post	Freedom from Pitch
Ash	A	B	C	C	A	C	C	A	A	A	A
Western red cedar	C	A	A	A	C	A	A	C	C	B	A
Cypress	B	B	B	A	B	A	B	B	B	B	A
Douglas-fir, larch	B	B	B-C	C	A	B	A	A	A	A	B
Gum	B	C	B	C	A	B	B	B	A	B	A
Hemlock, white fir[2]	B-C	B	B	C	C	C	B	B	A	B	A
Soft pines[3]	C	B	A	A	C	C	C	C	C	C	B
Southern pine	B	B	B	C	A	C	B	A	A	A	C
Poplar	C	B	B	A	B	C	C	B	B	B	A
Redwood	B	A	B	A	B	A	A	B	B	A	A
Spruce	C	A-B	B	B	B	C	C	B	B	B	A

[1] A—among the woods relatively high in the particular respect listed; B—among woods intermediate in that respect; C—among woods relatively low in that respect. Letters do not refer to lumber grades.

[2] Includes west coast and eastern hemlocks.

[3] Includes the western and northeastern pines.

WOODS—SPECIES AND TYPICAL USES

HARDWOODS

Name	Description	Hardness	Work-ability	Uses
Ash	Sapwood white, heartwood brown; strong and stiff.	Medium	Poor	Handles, vehicle parts, furniture, interior finish
Basswood	Sapwood creamy white, heartwood light brown; light in weight; straight grained	Soft	Good	Millwork, siding, kitchen cabinets, boxes, woodenware, furniture.
Beech	Sapwood whitish, heartwood reddish brown; heavy; broad and narrow wood rays or flecks, easily visible	Hard	Poor	Furniture, flooring, interior finish, handles, woodenware, containers.
Birch	Whitish sapwood, light to dark reddish brown heartwood; fine uniformed texture; heavy; takes beautiful natural finish	Hard	Poor	Interior finish, doors, furniture, flooring, woodenware, novelties.
Cherry	Sapwood nearly white, heartwood light to dark reddish brown with beautiful luster. Low in shrinkage.	Medium	Poor	Furniture, paneling, flooring, woodenware, caskets.
Gum	Sapwood flesh colored, heartwood of sweet gum is reddish brown, heartwood of other species light grey; moderately heavy; uniform texture.	Medium	Fair	Interior finish, millwork, containers, furniture (one of most used furniture woods for imitation of Walnut and Mahogany).
Hickory	Sapwood white, heartwood reddish brown; very tough; heavy; high shock resistance.	Hard	Poor	Tool handles, vehicle parts, sporting goods, agricultural implements, wheels.
Maple	Hard maple has white sapwood and light reddish brown heartwood; heavy; uniform texture. (Soft maple is lighter in color, less hard and strong.	Hard (Soft Maple Medium)	Poor	Flooring and other mill products; interior finish; handles, athletic equipment.
Oak	Sapwood off-white with brownish heartwood in white oak; red oak heartwood reddish brown; heavy; strong; pronounced flecks or wood rays visible on vertical grain lumber; pores large.	Hard	Poor	Flooring, interior finish, furniture, cabinet work, planing mill products. One of most used American woods.

Species	Description	Hardness	Working Quality	Uses
Walnut	Nearly white sapwood, light to dark chocolate brown heartwood; heavy; develops beautiful finish.	Medium	Fair	Cabinet work, furniture, interior finish, paneling, flooring, gun stocks.
Yellow Poplar	Sapwood white, heartwood yellowish brown with greenish tinge; light in weight; soft; straight grain and uniform texture.	Soft	Good	Siding, furniture, planing mill products, boxes, interior finish.

SOFTWOODS

Species	Description	Hardness	Working Quality	Uses
Cedar (Northern White)	Heartwood light brown to reddish tinge; sapwood light to white; light in weight; low in shrinkage; resistant to decay; aromatic odor.	Soft	Good	Siding, posts, pole ties, shingles, specialty purposes.
Cedar (Western Red)	Red to pink brown in color except sapwood, which is white; characteristic odor; highly resistant to decay.	Soft	Good	Siding, shingles, light construction, paneling, poles, posts, planing mill products, boxes.
Hemlock (Western)	Greylike white with purple tinge; annual rings distinct; no resin ducts; moderately light in weight; stiff	Medium	Fair to Good	General construction; planing mill products millwork, paneling, boxes and crates.
Larch (Western)	Russet to nut brown heartwood, finished surfaces appearing somewhat waxy; strong; sometimes confused with Douglas fir.	Hard	Poor	General construction, planing mill products, millwork, paneling, boxes and crates.
Pine (Lodgepole)	Sapwood almost white, heartwood slightly darker ranging from clear yellow to pale brown tinged with red; resinous odor and taste.	Soft	Good	Siding, framing, boards, paneling, ties, poles, mining timbers.
Pine (Ponderosa)	Creamy white to light reddish brown; narrow summerwood rings; uniform texture; moderately strong.	Soft	Good	Millwork and planing mill products, light construction, boxes, crates.
Pine (Southern)	Includes several species, yellow to reddish brown heartwood; resinous odor; distinct summerwood rings; moderately dense to dense; stiff; strong.	Hard	Poor	All types of building and general construction, siding, flooring, millwork, doors, boxes, crates.

Name	Description	Hardness	Work-ability	Uses
Pine (White)	Idaho, Northern white, Sugar pines are all genuine white pines; color creamy white to light brown; transition from springwood to summerwood gradual; soft uniform texture; light; stable under moisture changes.	Soft	Good	Millwork, planing mill products, sheathing and sub-flooring, boxes and crates.
Redwood	Dark, brown heartwood, creamy white sapwood; odorless; rings distinct; moderately light and strong.	Medium	Fair	General construction, planing mill products, tanks, vats, shingles, novelties.
Spruce (Sitka)	Heartwood yellow to pale brown with purplish cast, sapwood cream to light yellow in color; moderately light in weight; strong and stiff; high strength to wt. ratio.	Hard	Fair	Planing mill products, siding, millwork, finish, ladders, aircraft, boxes and crates.
Spruce (Engle-mann)	Pale off-white color; small light colored knots; fine grained; odorless; light in weight.	Soft	Fair	Planing mill products, siding, general construction, finish, scaffolding, industrial uses.

Appendix

ABBREVIATIONS—GENERAL

Air Entraining	A.E.	Flooring	Flg.	Properties	Prop.
Aggregate	Agg.	Foot—Feet	Ft. or ′	Per Square Foot	P.S.F.
Aluminum	Al.				
And	&	Galvanized	Galv.	Quantity	Quan.
Application	Appl.	Gauge	Ga.		
Architect	Arch.	Gravity	Gr.	Radius	Ra.
Asphalt	Asp.	Gypsum	Gyp.	Rating	Rtg.
Average	Av.			Recommend	Rec.
		Height	Ht.	Receivers	Rcvrs.
Barrel	Bbl.—Bbls.	Hexagonal	Hex.	Reducer	Red.
Basement	Bsmt.	High	Hi.	Reinforced	Reinf.
Board	Bd.	Hundred	C	Required	Reqd.
Brass	Br.			Roll	Rl.
Bundle	Bdl.—Bdls.	Inch	In. or ″	Roofing	Rfg.
		Inclusive	Inc.	Round	Rd.
Capacity	Cap.	Increaser	Incr.		
Casement	Csmt.	Inside Diameter	I.D.	Sack	Sk.
Cast Iron	C.I.	Insulating	Insul.	Sheathing	Shtg.
Carton	Ctn.			Sheet	Sht.—Shts.
Columns	Cols.	Joint	Jt.	Shingles	Shgls.
Commercial	Comm.			Siding	Sdg.
Circumference	Cir.	Length	Lth.	Single	Sgl.
Compound	Comp.	Light—Lights	Lt.—Lts.	Sliding	Sldg.
Conductivity	Cond.	Lining	Lng.	Smooth	Sm.
Construction	Const.	Liquid	Liq.	Specific Gravity	Sp. Gr.
Combination	Comb.	Load	Ld.	Square	Sq.
Concrete	Conc.			Standard	Std.
Copper	Copr.	Machine	Mach.	Steel	Stl.
Cubic	Cu.	Material	Matrl—Matrls.	Strength	Str.
		Metal	Mtl.	Surfaced 4 Sides	S4S
Diameter	Dia.	Mile	Mi.		
Dimension	Dim.			Temperature	Temp.
Ditto	″	Number	No.	Thousand	M
Double Hung	D.H.	Outside Diameter	O.D.	Tongue & Grooved	T & G
Double	Dbl.	Opening	Opng.—Opngs.	Total	Tot.
		Overall	O.A.	Transmission	Trans.
Estimate—ed	Est.-Estd.	On Center	O.C.		
Electrical	Elect.			Ventilator	Vent.
Equipment	Equipt.	Package	Pkg.	Vitrified	Vit.
Equivalent	Equiv.	Packed	Pkd.		
Exposure	Exp.	Panel	Panl.	Waterproofing	Wpfg.
		Paper	Ppr.	Width	Wth.
Finish	Fin.	Pieces	Pcs.		
		Plaster	Pl.	Yard	Yd.
		Pound	Lb.—Lbs.—#		

ABBREVIATIONS—LUMBER, STANDARD

AAR	Association of American Railroads	D2S&CM	dressed two sides and center matched
AD	air dried	D2S&SM	dressed two sides and standard matched
ADF	after deducting freight	E	edge
ALS	American Lumber Standard	EB1S	edge bead one side
AST	antistain treated. At ship tackle (western softwoods)	EB2S, SB2S	edge bead on two sides
AV or avg	average	EE	eased edges
AW&L	all widths and lengths	EG	edge (vertical or rift) grain
B1S	see EB1S, CB1S, and E&CB1S	EM	end matched
B2S	see EB2S, CB2S, and E&CB2S	EV1S, SV1S	edge V one side
B&B, B&BTR	B and Better	EV2S, SV2S	edge V two sides
B&S	beams and stringers	E&CB1S,	edge and center bead one side
BD	board	E&CB2S, DB2S, BC&2S	edge and center bead two sides
BD FT	board feet		
BDL	bundle		
BEV	bevel or beveled	E&CV1S, DV1S, V&CV1S	edge and center V one side
BH	boxed heart		
B/L,BL	bill of lading		
BM	board measure	E&CV2S, DV2S, V&CV2S	edge and center V two sides
BSND	bright sapwood no defect		
BTR	better		
c	allowable stress in compression in pounds per square inch	f	allowable stress in bending in pounds per square inch
CB	center beaded	FA	facial area
CB1S	center bead on one side	Fac	factory
CB2S	center bead on two sides	FAS	free alongside (vessel)
CC	cubical content	FAS	Firsts and Seconds
cft or cu. ft.	cubic foot or feet	FBM, Ft. BM	feet board measure
CF	cost and freight		
CIF	cost, insurance, and freight	FG	flat or slash grain
CIFE	cost, insurance, freight, and exchange	FJ	finger joint. End-jointed lumber using a finger joint configuration
CG2E	center groove on two edges		
C/L	carload	FLG, Flg	flooring
CLG	ceiling	FOB	free on board (named point)
CLR	clear	FOHC	free of heart center
CM	center matched	FOK	free of knots
Com	common	FRT, Frt	freight
CS	calking seam	FT, ft	foot or feet
CSG	casing	FT. SM	feet surface measure
CV	center V	G	girth
CV1S	center V on one side	GM	grade marked
CV2S	center V on two sides	G/R	grooved roofing
DB Clg	double beaded ceiling (E&CB1S)	HB, H.B.	hollow back
DB Part	double beaded partition (E&CB2S)	HEM	hemlock
DET	double end trimmed	Hrt	heart
DF	Douglas-fir	H&M	hit and miss
DIM	dimension	H or M	hit or miss
DKG	decking	IN, in.	inch or inches
D/S, DS, D/Sdg	drop siding	Ind	industrial
D1S, D2S	See S1S and S2S	J&P	joists and planks
D&M	dressed and matched	JTD	jointed
D&CM	dressed and center matched	KD	kiln dried
D&SM	dressed and standard matched	LBR, Lbr	lumber

LCL	less than carload	
LGR	longer	
LGTH	length	
Lft, Lf	lineal foot or feet	
LIN, Lin	lineal	
LL	longleaf	
LNG, Lng	lining	
M	thousand	
MBM, MBF, M.BM	thousand (feet) board measure	
MC, M.C.	moisture content	
MERCH, Merch	merchantable	
MG	medium grain or mixed grain	
MLDG, Mldg	molding	
Mft	thousand feet	
MSR	machine stress rated	
N	nosed	
NBM	net board measure	
No.	number	
N1E or N2E	nosed one or two edges	
Ord	order	
PAD	partially air dry	
PAR, Par	paragraph	
PART, Part	partition	
PAT, Pat	pattern	
Pcs.	pieces	
PE	plain end	
PET	precision end trimmed	
P&T	posts and timbers	
P1S, P2S	see S1S and S2S	
RDM	random	
REG, Reg	regular	
Rfg.	roofing	
RGH, Rgh	rough	
R/L, RL	random lengths	
R/W, RW	random widths	
RES	resawn	
SB1S	single bead one side	
SDG, Sdg	siding	
S-DRY	surfaced dry. Lumber 19 percent moisture content or less per American Lumber Standard for softwood	
SE	square edge	
SEL, Sel	select or select grade	
SE&S	square edge and sound	
SG	slash or flat grain	
S-GRN	surfaced green. Lumber unseasoned, in	

		excess of 19 percent moisture content softwood
SGSSND	Sapwood, gum spots and streaks, no defect	
SIT.SPR	Sitka spruce	
S/L, SL, S/Lap	shiplap	
STD.M	standard matched	
SM	surface measure	
Specs	specifications	
SQ	square	
SQRS	squares	
SR	stress rated	
STD, Std	standard	
Std.lgths.	standard lengths	
SSND	sap stain no defect (stained)	
STK	stock	
STPG	stepping	
STR, STRUCT	structural	
SYP	southern yellow pine	
S&E	side and edge (surfaced on)	
S1E	surfaced one edge	
S2E	surfaced two edges	
S1S	surfaced one side	
S2S	surfaced two sides	
S4S	surfaced four sides	
S1S&CM	surfaced one side and center matched	
S2S&CM	surfaced two sides and center matched	
S4S&CS	surfaced four sides and calking seam	
S1S1E	surfaced one side, one edge	
S1S2E	surfaced one side, two edges	
S2S1E	surfaced two sides, one edge	
S2S&SL	surfaced two sides and shiplapped	
S2S&SM	surfaced two sides and standard matched	
t	allowable stress in tension in pounds per square inch	
TBR	timber	
T&G	tongued and grooved	
VG	vertical (edge) grain	
V1S	see EV1S, CV1S, and E&CV1S	
V2S	see EV2S, CV2S, and E&CV2S	
WCH	west coast hemlock	
WDR, wdr	wider	
WHAD	worm holes a defect	
WHND	worm holes no defect	
WT	weight	
WTH	width	
WRD	western redcedar	
YP	yellow pine	

AMORTIZATION OF LOANS
Monthly Payment per $1000

Years	Interest Rates						
	8	8.5	9	9.5	10	10.5	11
1	86.99	87.22	87.46	87.69	87.92	88.15	88.39
2	45.23	45.46	45.69	45.92	46.15	46.38	46.61
3	31.34	31.57	31.80	32.04	32.27	32.51	32.74
4	24.42	24.65	24.89	25.13	25.37	25.61	25.85
5	20.28	20.52	20.76	21.01	21.25	21.50	21.75
6	17.54	17.78	18.03	18.28	18.53	18.78	19.04
7	15.59	15.84	16.09	16.35	16.61	16.87	17.13
8	14.14	14.40	14.66	14.92	15.18	15.45	15.71
9	13.02	13.28	13.55	13.81	14.08	14.36	14.63
10	12.14	12.40	12.67	12.94	13.22	13.50	13.78
15	9.56	9.85	10.15	10.45	10.75	11.06	11.37
20	8.37	8.68	9.00	9.33	9.66	9.99	10.33
25	7.72	8.06	8.40	8.74	9.09	9.45	9.81
30	7.34	7.69	8.05	8.41	8.78	9.15	9.53
35	7.11	7.47	7.84	8.22	8.60	8.99	9.37
40	6.96	7.34	7.72	8.11	8.50	8.89	9.29

ESTIMATING AIDS
This chart shows in alphabetical order building parts, components and items with the take-off unit required and the estimating unit generally used in the industry. Further data, information, and charts on items listed may be found by checking the index and referring to information under the listing indicated.

Item	Take-off Unit	Estimating Unit
A		
ACOUSTICAL	Sq. Ft.	C. Sq. Ft.
AIR COND.		
B		
BACKFILL	Cu. Yd.	Cu. Yds.
BAR JOIST	Size	Wt. Lin. Ft.
BEAMS		
Concrete	Lin. Ft.	C. Lin. Ft.
FiberGlass	" "	Lbs. Lin. Ft.
Steel	" "	Lbs. Lin. Ft.
Wood	" "	M. Bd. Ft.-L.F.
BRIDGING	Sets-L.F.	Sets-L.F.
C		
CHIMNEYS	Feet	Foot

Item	Take-off Unit	Estimating Unit
CLOSETS		
COLUMNS		
Concrete	Feet	Ft.
Steel	Units	Unit-Ft.
Wood	Ft.	L.F.-M.Bd. Ft.
COMPOSITION		
BOARD	Sq. Ft.	C. Sq. Ft.
D		
DOORS		
Cabinet	Units	Unit
Interior	"	"
Exterior	"	"
Garage	"	"
Frames	Units-Set	Units-Sets
DRAIN TILE	Lin. Ft.	C. Lin. Ft.
DRIVEWAYS	Sq. Ft.	C. Sq. Ft.
DRYWALL	Sq. Ft.	C. Sq. Ft.
E		
ELECTRICAL	Sub. Bid.	Job.
EXCAVATING	Cu. Yds.	Cu. Yds.
F		
FIREPLACE	Units	Unit
FIBERGLASS	Sq. Ft.	Sq. Ft.
FLOORS		
Carpet	Sq. Yds.	Sq. Yd.
Ceramic Tile	Sq. Ft.	C. Sq. Ft.
Concrete	" "	C. Sq. Ft.-Cu. Yd.
Resilient Tile/Sheet	" "	" " " "
Wood	" "	C. Sq. Ft.-M.Bd.F.
FOOTERS	Lin. Ft.-Cu. Ft.	C. Lin. Ft.-Cu. Yds.
FOUNDATIONS		
Concrete	Cu. Ft.-Cu. Yds.	Cu. Yds.
FRAMING		
Beams—		
Columns	Lin. Ft. Units	Lin. Ft.-M. Bd. Ft.-Unit
Joists	Sq. Ft. PCS	M. Bd. F-Lin. F.
Rafters	" " "	M. Bds. Ft.-Units
Sills	Lin. Ft.-PSC	M. Bd. Ft.
Studs	S. F.-L. F. PCS	M. Bd. Ft.
G		
GARAGE		
GLASS	Pcs.-Sq. Ft.	Units-Sq. Ft.
GLASS BLOCK	Sq. Ft.	M. Units
GUTTERS—		
CONDUCTOR	Lin. Ft.	Ft.

Item	Take-off Unit	Estimating Unit
H		
HARDBOARD	Sq. Ft.	C.S.F.-Panels
HEATING	Sub. Bid.	Job.
I		
I BEAMS	Lin. Ft.	Lin. Ft.-Wt.
INSULATION		
Rigid	Sq. Ft.	C. Sq. Ft.
Pouring	" "	C.S.F.-Cu. Ft.
Panels	" "	C.S.F.-Pcs.
Reflective	" "	C. Sq. Ft.
Blanket	" "	" " "
L		
LATH		
Gypsum	Sq. Ft.	M.S. Ft. Bd.
Lumber	Sq. Ft.	M. Bd. Ft.
Metal	" "	C.S.F.-Sq. Yds.
Lintels	Lin. Ft. Units	L.F.-Unit-wt.
M		
MASONRY	Sq. Ft.	various
Brick	" "	Per M
Concrete	" "	M or Units
Stone	" "	C. Sq. Ft.
N		
NAILS	Sq. Ft.	Lbs.
P		
PAINTING	Sq. Ft.	C. Sq. Ft.
PAPERING	" "	Sq. Ft. Yds.-Roll
PARTITIONS	" "	Sq. Ft.
PERIMETER INSUL.	Lin. Ft.	Lin. Ft.-Sq. Ft.
PATIOS	Sq. Ft.	Sq. Ft.-M-Pcs.
PLASTER	Sq. Ft.	C. Sq. Yds.
PLASTIC FOAM	Sq. Ft.	Sq. Ft. Pc.
PLASTIC SHEET	" "	" " "
PLUMBING	Sub. Bid.	Job.
PLYWOOD	Sq. Ft.	M.S.F.-Pc.
POSTS	Unit	Each
R		
REINFORCING	Lin. Ft.	Lin. Ft.-Wt.
ROOFING (All)	C. Sq. Ft.	Square
Asphalt Roll	" " "	"
Asphalt Shingles	" " "	"
Built-Up	" " "	"
Metal	" " "	"

Item	Take-off Unit	Estimating Unit
Slate	" " "	"
Tile	" " "	"
Wood Shingles	" " "	"
ROOM AREAS	Sq. Ft.	Sq. Ft.
S		
SEPTIC SYSTEMS	Unit	Unit
SEWERS	Lin. Ft.	C.L.F.-Jts.
SHEATHING		
Wood	Sq. Ft.	M. Bd. Ft.
Gypsum	" "	M. Sq. Ft.
Insulating		M. Sq. Ft. Pnls
Plywood	" "	" " " "
SIDING		
Aluminum	Sq. Ft.	C or M.S.F.-Sq.
Steel	" "	" " " "
Vinyl	" "	" " " "
Wood	" "	M. Bd. Ft.
Wood Shingles	" "	C or M.S.F.-Sq.
SOFFITS	" "	L.F.-Sq. Ft.
STAIRS	Units	Unit
STEEL		
Angles	Units-Ft.	Unit-Ft. Wt.
Beams	" "	" " "
STEPS	Units	Unit
STUCCO	Sq. Ft.	C. Sq. Ft.
T		
TILE		
Resilient-Parquet	Sq. Ft.	C. Sq. Ft.
Ceramic	" "	" " "
Metal	" "	" " "
Plastic	" "	" " "
TIMBER	Lin. Ft.	C. Bd. Ft.
TRENCHING	Lin. Ft.	C. Lin. Ft.
TRIM INT.	Lin. Ft.-Unit-Set	Lin Ft.-Unit-Sets
TRIM EXT.	Lin. Ft.	Bd. Ft.-Unit-L.F.
U		
UNDERLAYMENT	Sq. Ft.	C. Sq. Ft.
V		
VAPOR BARRIER	Sq. Ft.	C. Sq. Ft.
VENTILATION	Cu. Ft.	Unit
W		
WALKS	Sq. Ft.	C. Sq. Ft.
WALLS INT.	Sq. Ft.	C. Sq. Ft.
WATERPROOFING	Sq. Ft.	C. Sq. Ft.
WINDOWS	Units	Unit

ESTIMATING CHECKLIST

Excavating	Insulation	Porches and Steps
Footing	Roof Framing	Finish Flooring
Drain Tile	Ceiling Insulation	Staircases
Sewer	Roof Sheathing	Interior Trim
Foundation	Roofing	Doors
Waterproofing	Flashing	Hardware
Backfill	Windows	Cabinets-Built-ins
Porch Footings	Chimney	Tile
Chimney Footings	Fireplace/Wood Stove	Paneling
Posts and Beams	Plumbing-Rough	Interior Painting
1st Floor Joists	Plumbing-Finish	Wallcoverings
Subfloor	Electric Wiring	Interior Decorating
Studding	Heating-Rough	Electric Fixtures
Door and Wdw. Fr.	Heating Unit and Rad.	Electric Equipment
Exterior Walls	Air Conditioning	Appliances
Partition Framing	Lathing	Grading
Sheathing	Plastering	Landscaping
Insulation	Drywall	Patios
2nd Floor Joists	Glazing	Caulk W. Str. and Clean
Subfloor	Gutters & Downspouts	Screens and St. Sash
Studding	Exterior Painting	Garage Floor
Exterior Walls	Basement Floor	Garage Walls
Partition Framing	Concrete Walks & Drives	Garage Roof
Sheathing	Decks	Garage Doors & Windows

AMORTIZATION OF LOANS
Monthly Payment per $1000

Year	Interest Rates 11.5	12	12.5	13	13.5	14	14.5
1	88.62	88.85	89.09	89.32	89.56	89.79	90.03
2	46.85	47.08	47.31	47.55	47.78	48.02	48.25
3	32.98	33.22	33.46	33.70	33.94	34.18	34.43
4	26.09	26.34	26.58	26.83	27.08	27.33	27.58
5	22.00	22.25	22.50	22.76	23.01	23.27	23.53
6	19.30	19.56	19.82	20.08	20.34	20.61	20.88
7	17.39	17.66	17.93	18.20	18.47	18.75	19.02
8	15.98	16.26	16.53	16.81	17.09	17.38	17.66
9	14.91	15.19	15.47	15.76	16.05	16.34	16.63
10	14.06	14.35	14.64	14.94	15.23	15.53	15.83
15	11.69	12.01	12.33	12.66	12.99	13.32	13.66
20	10.67	11.02	11.37	11.72	12.08	12.44	12.80
25	10.17	10.54	10.91	11.28	11.66	12.04	12.43
30	9.91	10.29	10.68	11.07	11.46	11.85	12.25
35	9.77	10.16	10.56	10.96	11.36	11.76	12.17
40	9.69	10.09	10.49	10.90	11.31	11.72	12.13

AMORTIZATION OF LOANS
Monthly Payment per $1000

Years	Interest Rates 15	15.5	16	16.5	17	17.5	18
1	90.26	90.50	90.74	90.97	91.21	91.44	91.68
2	48.49	48.73	48.97	49.21	49.45	49.68	49.92
3	34.67	34.92	35.16	35.41	35.66	35.90	36.15
4	27.84	28.09	28.35	28.60	28.86	29.11	29.38
5	23.79	24.06	24.32	24.59	24.86	25.12	25.39
6	21.15	21.42	21.70	21.97	22.25	22.53	22.81
7	19.30	19.58	19.87	20.15	20.44	20.73	21.02
8	17.95	18.24	18.53	18.83	19.13	19.42	19.72
9	16.93	17.23	17.53	17.83	18.14	18.45	18.76
10	16.14	16.45	16.76	17.07	17.38	17.70	18.02
15	14.00	14.34	14.69	15.04	15.40	15.75	16.10
20	13.17	13.54	13.92	14.29	14.67	15.05	15.43
25	12.81	13.20	13.59	13.99	14.38	14.78	15.17
30	12.65	13.05	13.45	13.86	14.26	14.66	15.07
35	12.57	12.98	13.39	13.80	14.21	14.62	15.03
40	12.54	12.95	13.36	13.77	14.19	14.60	15.01

ROOM AREAS
Total Square Foot Area 4 Walls and Ceiling—Deduct for All Openings

Ceilings 7 Feet High

	3'	4'	5'	6'	7'	8'	9'	10'	11'	12'	13'	14'	15'	16'	17'	18'	19'	20'	21'	22'
3'	93	110	127	144	161	178	195	212	229	246	263	280	297	314	331	348	365	382	399	416
4'	110	128	146	164	182	200	218	236	254	272	290	308	326	344	362	380	398	416	434	452
5'	127	146	165	184	203	222	241	260	279	298	317	336	355	374	393	412	431	450	469	488
6'	144	164	184	204	224	244	264	284	304	324	344	364	384	404	424	444	464	484	504	524
7'	161	182	203	224	245	266	287	308	329	350	371	392	413	434	455	476	497	518	539	560
8'	178	200	222	244	266	288	310	332	354	376	398	420	442	464	486	508	530	552	574	596
9'	195	218	241	264	287	310	333	356	379	402	425	448	471	494	517	540	563	586	609	632
10'	212	236	260	284	308	332	356	380	404	428	452	476	500	524	548	572	596	620	644	668
11'	229	254	279	304	329	354	379	404	429	454	479	504	529	554	579	604	629	654	679	704
12'	246	272	298	324	350	376	402	428	454	480	506	532	558	584	610	636	662	688	714	740
13'	263	290	317	344	371	398	425	452	479	506	533	560	587	614	641	668	695	722	749	776
14'	280	308	336	364	392	420	448	476	504	532	560	588	616	644	672	700	728	756	784	812
15'	297	326	355	384	413	442	471	500	529	558	587	616	645	674	703	732	761	790	819	848
16'	314	344	374	404	434	464	494	524	554	584	614	644	674	704	734	764	794	824	854	884
17'	331	362	393	424	455	486	517	548	579	610	641	672	703	734	765	796	827	858	889	920
18'	348	380	412	444	476	508	540	572	604	636	668	700	732	764	796	828	860	892	924	956
19'	365	398	431	464	497	530	563	596	629	662	695	728	761	794	827	860	893	926	959	992
20'	382	416	450	484	518	552	586	620	654	688	722	756	790	824	858	892	926	960	994	1028
21'	399	434	469	504	539	574	609	644	679	714	749	784	819	854	889	924	959	994	1029	1064
22'	416	452	488	524	560	596	632	668	704	740	776	812	848	884	920	956	992	1028	1064	1100
23'	433	470	507	544	581	618	655	692	729	766	803	840	877	914	951	988	1025	1062	1099	1136
24'	450	488	526	564	602	640	678	716	754	792	830	868	906	944	982	1020	1058	1096	1134	1172

ROOM AREAS

Total Square Foot Area 4 Walls and Ceiling—Deduct for All Openings

Ceilings 7 Feet 6 Inches High

	3'	4'	5'	6'	7'	8'	9'	10'	11'	12'	13'	14'	15'	16'	17'	18'	19'	20'	21'	22'
3'	99	117	135	153	171	189	207	225	243	261	279	297	315	333	351	369	387	405	423	441
4'	117	136	155	174	193	212	231	250	269	288	307	326	345	364	383	402	421	440	459	478
5'	135	155	175	195	215	235	255	275	295	315	335	355	375	395	415	435	455	475	495	515
6'	153	174	195	216	237	258	279	300	321	342	363	384	405	426	447	468	489	510	531	552
7'	171	193	215	237	259	281	303	325	347	369	391	413	435	457	479	501	523	545	567	589
8'	189	212	235	258	281	304	327	350	373	396	419	442	465	488	511	534	557	580	603	626
9'	207	231	255	279	303	327	351	375	399	423	447	471	495	519	543	567	591	615	639	663
10'	225	250	275	300	325	350	375	400	425	450	475	500	525	550	575	600	625	650	675	700
11'	243	269	295	321	347	373	399	425	451	477	503	529	555	581	607	633	659	685	711	737
12'	261	288	315	342	369	396	423	450	477	504	531	558	585	612	639	666	693	720	747	774
13'	279	307	335	363	391	419	447	475	503	531	559	587	615	643	671	699	727	755	783	811
14'	297	326	355	384	413	442	471	500	529	558	587	616	645	674	703	732	761	790	819	848
15'	315	345	375	405	435	465	495	525	555	585	615	645	675	705	735	765	795	825	855	885
16'	333	364	395	426	457	488	519	550	581	612	643	674	705	736	767	798	829	860	891	922
17'	351	383	415	447	479	511	543	575	607	639	671	703	735	767	799	831	863	895	927	959
18'	369	402	435	468	501	534	567	600	633	666	699	732	765	798	831	864	897	930	963	996
19'	387	421	455	489	523	557	591	625	659	693	727	761	795	829	863	897	931	965	999	1033
20'	405	440	475	510	545	580	615	650	685	720	755	790	825	860	895	930	965	1000	1035	1070
21'	423	459	495	531	567	603	639	675	711	747	783	819	855	891	927	963	999	1035	1071	1107
22'	441	478	515	552	589	626	663	700	737	774	811	848	885	922	959	996	1033	1070	1107	1144
23'	459	497	535	573	611	649	687	725	763	801	839	877	915	953	991	1029	1067	1105	1143	1181
24'	477	516	555	594	633	672	711	750	789	828	867	906	945	984	1023	1062	1101	1140	1179	1218

ROOM AREAS

Total Square Foot Area 4 Walls and Ceiling—Deduct for All Openings

Ceilings 8 Feet High

	3'	4'	5'	6'	7'	8'	9'	10'	11'	12'	13'	14'	15'	16'	17'	18'	19'	20'	21'	22'
3'	105	124	143	162	181	200	219	238	257	276	295	314	333	352	371	390	409	428	447	466
4'	124	144	164	184	204	224	244	264	284	304	324	344	364	384	404	424	444	464	484	504
5'	143	164	185	206	227	248	269	290	311	332	353	374	395	416	437	458	479	500	521	542
6'	162	184	206	228	250	272	294	316	338	360	382	404	426	448	470	492	514	536	558	580
7'	181	204	227	250	273	296	319	342	365	388	411	434	457	480	503	526	549	572	595	618
8'	200	224	248	272	296	320	344	368	392	416	440	464	488	512	536	560	584	608	632	656
9'	219	244	269	294	319	344	369	394	419	444	469	494	519	544	569	594	619	644	669	694
10'	238	264	290	316	342	368	394	420	446	472	498	524	550	576	602	628	654	680	706	732
11'	257	284	311	338	365	392	419	446	473	500	527	554	581	608	635	662	689	716	743	770
12'	276	304	332	360	388	416	444	472	500	528	556	584	612	640	668	696	724	752	780	808
13'	295	324	353	382	411	440	469	498	527	556	585	614	643	672	701	730	759	788	817	846
14'	314	344	374	404	434	464	494	524	554	584	614	644	674	704	734	764	794	824	854	884
15'	333	364	395	426	457	488	519	550	581	612	643	674	705	736	767	798	829	860	891	922
16'	352	384	416	448	480	512	544	576	608	640	672	704	736	768	800	832	864	896	928	960
17'	371	404	437	470	503	536	569	602	635	668	701	734	767	800	833	866	899	932	965	998
18'	390	424	458	492	526	560	594	628	662	696	730	764	798	832	866	900	934	968	1002	1036
19'	409	444	479	514	549	584	619	654	689	724	759	794	829	864	899	934	969	1004	1039	1074
20'	428	464	500	536	572	608	644	680	716	752	788	824	860	896	932	968	1004	1040	1076	1112
21'	447	484	521	558	595	632	669	706	743	780	817	854	891	928	965	1002	1039	1076	1113	1150
22'	466	504	542	580	618	656	694	732	770	808	846	884	922	960	998	1036	1074	1112	1150	1188
23'	485	524	563	602	641	680	719	758	797	836	875	914	953	992	1031	1070	1109	1148	1187	1226
24'	504	544	584	624	664	704	744	784	824	864	904	944	984	1024	1064	1104	1144	1184	1224	1264

ROOM AREAS

Total Square Foot Area 4 Walls and Ceiling—Deduct for All Openings

Ceilings 8 Feet 6 Inches High

	3'	4'	5'	6'	7'	8'	9'	10'	11'	12'	13'	14'	15'	16'	17'	18'	19'	20'	21'	22'
3'	111	131	151	171	191	211	231	251	271	291	311	331	351	371	391	411	431	451	471	491
4'	131	152	173	194	215	236	257	278	299	320	341	362	383	404	425	446	467	488	509	530
5'	151	173	195	217	239	261	283	305	327	349	371	393	415	437	459	481	503	525	547	569
6'	171	194	217	240	263	286	309	332	355	378	401	424	447	470	493	516	539	562	585	608
7'	191	215	239	263	287	311	335	359	383	407	431	455	479	503	527	551	575	599	623	647
8'	211	236	261	286	311	336	361	386	411	436	461	486	511	536	561	586	611	636	661	686
9'	231	257	283	309	335	361	387	413	439	465	491	517	543	569	595	621	647	673	699	725
10'	251	278	305	332	359	386	413	440	467	494	521	548	575	602	629	656	683	710	737	764
11'	271	299	327	355	383	411	439	467	495	523	551	579	607	635	663	691	719	747	775	803
12'	291	320	349	378	407	436	465	494	523	552	581	610	639	668	697	726	755	784	813	842
13'	311	341	371	401	431	461	491	521	551	581	611	641	671	701	731	761	791	821	851	881
14'	331	362	393	424	455	486	517	548	579	610	641	672	703	734	765	796	827	858	889	920
15'	351	383	415	447	479	511	543	575	607	639	671	703	735	767	799	831	863	895	927	959
16'	371	404	437	470	503	536	569	602	635	668	701	734	767	800	833	866	899	932	965	998
17'	391	425	459	493	527	561	595	629	663	697	731	765	799	833	867	901	935	969	1003	1037
18'	411	446	481	516	551	586	621	656	691	726	761	796	831	866	901	936	971	1006	1041	1076
19'	431	467	503	539	575	611	647	683	719	755	791	827	863	899	935	971	1007	1043	1079	1115
20'	451	488	525	562	599	636	673	710	747	784	821	858	895	932	969	1007	1043	1080	1117	1154
21'	471	509	547	585	623	661	699	737	775	813	851	889	927	965	1003	1041	1079	1117	1155	1193
22'	491	530	569	608	647	686	725	764	803	842	881	920	959	998	1037	1076	1115	1154	1193	1232
23'	511	551	591	631	671	711	751	791	831	871	911	951	991	1031	1071	1111	1151	1191	1231	1271
24'	531	572	613	654	695	736	777	818	859	900	941	982	1023	1064	1105	1146	1187	1228	1269	1310

ROOM AREAS

Total Square Foot Area 4 Walls and Ceiling—Deduct for All Openings

Ceilings 9 Feet High

	3'	4'	5'	6'	7'	8'	9'	10'	11'	12'	13'	14'	15'	16'	17'	18'	19'	20'	21'	22'
3'	117	138	159	180	201	222	243	264	285	306	327	348	369	390	411	432	453	474	495	516
4'	138	160	182	204	226	248	270	292	314	336	358	380	402	424	446	468	490	512	534	556
5'	159	182	205	228	251	274	297	320	343	366	389	412	435	458	481	504	527	550	573	596
6'	180	204	228	252	276	300	324	348	372	396	420	444	468	492	516	540	564	588	612	636
7'	201	226	251	276	301	326	351	376	401	426	451	476	501	526	551	576	601	626	651	676
8'	222	248	274	300	326	352	378	404	430	456	482	508	534	560	586	612	638	664	690	716
9'	243	270	297	324	351	378	405	432	459	486	513	540	567	594	621	648	675	702	729	756
10'	264	292	320	348	376	404	432	460	488	516	544	572	600	628	656	684	712	740	768	796
11'	285	314	343	372	401	430	459	488	517	546	575	604	633	662	691	720	749	778	807	836
12'	306	336	366	396	426	456	486	516	546	576	606	636	666	696	726	756	786	816	846	876
13'	327	358	389	420	451	482	513	544	575	606	637	668	699	730	761	792	823	854	885	916
14'	348	380	412	444	476	508	540	572	604	636	668	700	732	764	796	828	860	892	924	956
15'	369	402	435	468	501	534	567	600	633	666	699	732	765	798	831	864	897	930	963	996
16'	390	424	458	492	526	560	594	628	662	696	730	764	798	832	866	900	934	968	1002	1036
17'	411	446	481	516	551	586	621	656	691	726	761	796	831	866	901	936	971	1006	1041	1076
18'	432	468	504	540	576	612	648	684	720	756	792	828	864	900	936	972	1008	1044	1080	1116
19'	453	490	527	564	601	638	675	712	749	786	823	860	897	934	971	1008	1045	1082	1119	1156
20'	474	512	550	588	626	664	702	740	778	816	854	892	930	968	1006	1044	1082	1120	1158	1196
21'	495	534	573	612	651	690	729	768	807	846	885	924	963	1002	1041	1080	1119	1158	1197	1236
22'	516	556	596	636	676	716	756	796	836	876	916	956	996	1036	1076	1116	1156	1196	1236	1276
23'	537	578	619	660	701	742	783	824	865	906	947	988	1029	1070	1111	1152	1193	1234	1275	1316
24'	558	600	642	684	726	768	810	852	894	936	978	1020	1062	1104	1146	1188	1230	1272	1314	1356

ROOM AREAS
Total Square Foot Area 4 Walls and Ceiling—Deduct for All Openings

Ceilings 9 Feet 6 Inches High

	3'	4'	5'	6'	7'	8'	9'	10'	11'	12'	13'	14'	15'	16'	17'	18'	19'	20'	21'	22'
3'	123	145	167	189	211	233	255	277	299	321	343	365	387	409	431	453	475	497	519	541
4'	145	168	191	214	237	260	283	306	329	352	375	398	421	444	467	490	513	536	559	582
5'	167	191	215	239	263	287	311	335	359	383	407	431	455	479	503	527	551	575	599	623
6'	189	214	239	264	289	314	339	364	389	414	439	464	489	514	539	564	589	614	639	664
7'	211	237	263	289	315	341	367	393	419	445	471	497	523	549	575	601	627	653	679	705
8'	233	260	287	314	341	368	395	422	449	476	503	530	557	584	611	638	665	692	719	746
9'	255	283	311	339	367	395	423	451	479	507	535	563	591	619	647	675	703	731	759	787
10'	277	306	335	364	393	422	451	480	509	538	567	596	625	654	683	712	741	770	799	828
11'	299	329	359	389	419	449	479	509	539	569	599	629	659	689	719	749	779	809	839	869
12'	321	352	383	414	445	476	507	538	569	600	631	662	693	724	755	786	817	848	879	910
13'	343	375	407	439	471	503	535	567	599	631	663	695	727	759	791	823	855	887	919	951
14'	365	398	431	464	497	530	563	596	629	662	695	728	761	794	827	860	893	926	959	992
15'	387	421	455	489	523	557	591	625	659	693	727	761	795	829	863	897	931	965	999	1033
16'	409	444	479	514	549	584	619	654	689	724	759	794	829	864	899	934	969	1004	1039	1074
17'	431	467	503	539	575	611	647	683	719	755	791	827	863	899	935	971	1007	1043	1079	1115
18'	453	490	527	564	601	638	675	712	749	786	823	860	897	934	971	1008	1045	1082	1119	1156
19'	475	513	551	589	627	665	703	741	779	817	855	893	931	969	1007	1045	1083	1121	1159	1197
20'	497	536	575	614	653	692	731	770	809	848	887	926	965	1004	1043	1082	1121	1160	1199	1238
21'	519	559	599	639	679	719	759	799	839	879	919	959	999	1039	1079	1119	1159	1199	1239	1279
22'	541	582	623	664	705	746	787	828	869	910	951	992	1033	1074	1115	1156	1197	1238	1279	1320
23'	563	605	647	689	731	773	815	857	899	941	983	1025	1067	1109	1151	1193	1235	1277	1319	1361
24'	585	628	671	714	757	800	843	886	929	972	1015	1058	1001	1144	1187	1230	1273	1316	1359	1402

ROOM AREAS

Total Square Foot Area 4 Walls and Ceiling—Deduct for All Openings

Ceilings 10 Feet High

	3'	4'	5'	6'	7'	8'	9'	10'	11'	12'	13'	14'	15'	16'	17'	18'	19'	20'	21'	22'
3'	129	152	175	198	221	244	267	290	313	336	359	382	405	428	451	474	497	520	543	566
4'	152	176	200	224	248	272	296	320	344	368	392	416	440	464	488	512	536	560	584	608
5'	175	200	225	250	275	300	325	350	375	400	425	450	475	500	525	550	575	600	625	650
6'	198	224	250	276	302	328	354	380	406	432	458	484	510	536	562	588	614	640	666	692
7'	221	248	275	302	329	356	383	410	437	464	491	518	545	572	599	626	653	680	707	734
8'	244	272	300	328	356	384	412	440	468	496	524	552	580	608	636	664	692	720	748	776
9'	267	296	325	354	383	412	441	470	499	528	557	586	615	644	673	702	731	760	789	818
10'	290	320	350	380	410	440	470	500	530	560	590	620	650	680	710	740	770	800	830	860
11'	313	344	375	406	437	468	499	530	561	592	623	654	685	716	747	778	809	840	871	902
12'	336	368	400	432	464	496	528	560	592	624	656	688	720	752	784	816	848	880	912	944
13'	359	392	425	458	491	524	557	590	623	656	689	722	755	788	821	854	887	920	953	986
14'	382	416	450	484	518	552	586	620	654	688	722	756	790	824	858	892	926	960	994	1028
15'	405	440	475	510	545	580	615	650	685	720	755	790	825	860	895	930	965	1000	1035	1070
16'	428	464	500	536	572	608	644	680	716	752	788	824	860	896	932	968	1004	1040	1076	1112
17'	451	488	525	562	599	636	673	710	747	784	821	858	895	932	969	1006	1043	1080	1117	1154
18'	474	512	550	588	626	664	702	740	778	816	854	892	930	968	1006	1044	1082	1120	1158	1196
19'	497	536	575	614	653	692	731	770	809	848	887	926	965	1004	1043	1082	1121	1160	1199	1238
20'	520	560	600	640	680	720	760	800	840	880	920	960	1000	1040	1080	1120	1160	1200	1240	1280
21'	543	584	625	666	707	748	789	830	871	912	953	994	1035	1076	1117	1158	1199	1240	1281	1322
22'	566	608	650	692	734	776	818	860	902	944	986	1028	1070	1112	1154	1196	1238	1280	1322	1364
23'	589	632	675	718	761	804	847	890	933	976	1019	1062	1105	1148	1191	1234	1277	1320	1363	1406
24'	612	656	700	744	788	832	876	920	964	1008	1052	1096	1140	1184	1228	1272	1316	1360	1404	1448

ROOM AREAS

Total Square Foot Area 4 Walls and Ceiling—Deduct for All Openings

Ceilings 10 Feet 6 Inches High

	3'	4'	5'	6'	7'	8'	9'	10'	11'	12'	13'	14'	15'	16'	17'	18'	19'	20'	21'	22'
3'	135	159	183	207	231	255	279	303	327	351	375	399	423	447	471	495	519	543	567	591
4'	159	184	209	234	259	284	309	334	359	384	409	434	459	484	509	534	559	584	609	634
5'	183	209	235	261	287	313	339	365	391	417	443	469	495	521	547	573	599	625	651	677
6'	207	234	261	288	315	342	369	396	423	450	477	504	531	558	585	612	639	666	693	720
7'	231	259	287	315	343	371	399	427	455	483	511	539	567	595	623	651	679	707	735	763
8'	255	284	313	342	371	400	429	458	487	516	545	574	603	632	661	690	719	748	777	806
9'	279	309	339	369	399	429	459	489	519	549	579	609	639	669	699	729	759	789	819	849
10'	303	334	365	396	427	458	489	520	551	582	613	644	675	706	737	768	799	830	861	892
11'	327	359	391	423	455	487	519	551	583	615	647	679	711	743	775	807	839	871	903	935
12'	351	384	417	450	483	516	549	582	615	648	681	714	747	780	813	846	879	912	945	978
13'	375	409	443	477	511	545	579	613	647	681	715	749	783	817	851	885	919	953	987	1021
14'	399	434	469	504	539	574	609	644	679	714	749	784	819	854	889	924	959	994	1029	1064
15'	423	459	495	531	567	603	639	675	711	747	783	819	855	891	927	963	999	1035	1071	1107
16'	447	484	521	558	595	632	669	706	743	780	817	854	891	928	965	1002	1039	1076	1113	1150
17'	471	509	547	585	623	661	699	737	775	813	851	889	927	965	1003	1041	1079	1117	1155	1193
18'	495	534	573	612	651	690	729	768	807	846	885	924	963	1002	1041	1080	1119	1158	1197	1236
19'	519	559	599	639	679	719	759	799	839	879	919	959	999	1039	1079	1119	1159	1199	1239	1279
20'	543	584	625	666	707	748	789	830	871	912	953	994	1035	1076	1117	1158	1199	1240	1281	1322
21'	567	609	651	693	735	777	819	861	903	945	987	1029	1071	1113	1155	1197	1239	1281	1323	1365
22'	591	634	677	720	763	806	849	892	935	978	1021	1064	1107	1150	1193	1236	1279	1322	1365	1408
23'	615	659	703	747	791	835	879	923	967	1011	1055	1099	1143	1187	1231	1275	1319	1363	1407	1451
24'	639	684	729	774	819	864	909	954	999	1044	1089	1134	1179	1224	1269	1314	1359	1404	1449	1494

ROOM AREAS
Total Square Foot Area 4 Walls and Ceiling—Deduct for All Openings

Ceilings 11 Feet High

	3'	4'	5'	6'	7'	8'	9'	10'	11'	12'	13'	14'	15'	16'	17'	18'	19'	20'	21'	22'
3'	141	166	191	216	241	266	291	316	341	366	391	416	441	466	491	516	541	566	591	616
4'	166	192	218	244	270	296	322	348	374	400	426	452	478	504	530	556	582	608	634	660
5'	191	218	245	272	299	326	353	380	407	434	461	488	515	542	569	596	623	650	677	704
6'	216	244	272	300	328	356	384	412	440	468	496	524	552	580	608	636	664	692	720	748
7'	241	270	299	328	357	386	415	444	473	502	531	560	589	618	647	676	705	734	763	792
8'	266	296	326	356	386	416	446	476	506	536	566	596	626	656	686	716	746	776	806	836
9'	291	322	353	384	415	446	477	508	539	570	601	632	663	694	725	756	787	818	849	880
10'	316	348	380	412	444	476	508	540	572	604	636	668	700	732	764	796	828	860	892	924
11'	341	374	407	440	473	506	539	572	605	638	671	704	737	770	803	836	869	902	935	968
12'	366	400	434	468	502	536	570	604	638	672	706	740	774	808	842	876	910	944	978	1012
13'	391	426	461	496	531	566	601	636	671	706	741	776	811	846	881	916	951	986	1021	1056
14'	416	452	488	524	560	596	632	668	704	740	776	812	848	884	920	956	992	1028	1064	1100
15'	441	478	515	552	589	626	663	700	737	774	811	848	885	922	959	996	1033	1070	1107	1144
16'	466	504	542	580	618	656	694	732	770	808	846	884	922	960	998	1036	1074	1112	1150	1188
17'	491	530	569	608	647	686	725	764	803	842	881	920	959	998	1037	1076	1115	1154	1193	1232
18'	516	556	596	636	676	716	756	796	836	876	916	956	996	1036	1076	1116	1156	1196	1236	1276
19'	541	582	623	664	705	746	787	828	869	910	951	992	1033	1074	1115	1156	1197	1238	1279	1320
20'	566	608	650	692	734	776	818	860	902	944	986	1028	1070	1112	1154	1196	1238	1280	1322	1364
21'	591	634	677	720	763	806	849	892	935	978	1021	1064	1107	1150	1193	1236	1279	1322	1365	1408
22'	616	660	704	748	792	836	880	924	968	1012	1056	1100	1144	1188	1232	1276	1320	1364	1408	1452
23'	641	686	731	776	821	866	911	956	1001	1046	1091	1136	1181	1226	1271	1316	1361	1406	1451	1496
24'	666	712	758	804	850	896	942	988	1034	1080	1126	1172	1218	1264	1310	1356	1402	1448	1494	1540

ROOM AREAS

Total Square Foot Area 4 Walls and Ceiling—Deduct for All Openings

Ceilings 12 Feet High

	3'	4'	5'	6'	7'	8'	9'	10'	11'	12'	13'	14'	15'	16'	17'	18'	19'	20'	21'	22'
3'	153	180	207	234	261	288	315	342	369	396	423	450	477	504	531	558	585	612	639	666
4'	180	208	236	264	292	320	348	376	404	432	460	488	516	544	572	600	628	656	684	712
5'	207	236	265	294	323	352	381	410	439	468	497	526	555	584	613	642	671	700	729	758
6'	234	264	294	324	354	384	414	444	474	504	534	564	594	624	654	684	714	744	774	804
7'	261	292	323	354	385	416	447	478	509	540	571	602	633	664	695	726	757	788	819	850
8'	288	320	352	384	416	448	480	512	544	576	608	640	672	704	736	768	800	832	864	896
9'	315	348	381	414	447	480	513	546	579	612	645	678	711	744	777	810	843	876	909	942
10'	342	376	410	444	478	512	546	580	614	648	682	716	750	784	818	852	886	920	954	988
11'	369	404	439	474	509	544	579	614	649	684	719	754	789	824	859	894	929	964	999	1034
12'	396	432	468	504	540	576	612	648	684	720	756	792	828	864	900	936	972	1008	1044	1080
13'	423	460	497	534	571	608	645	682	719	756	793	830	867	904	941	978	1015	1052	1089	1126
14'	450	488	526	564	602	640	678	716	754	792	830	868	906	944	982	1020	1058	1096	1134	1172
15'	477	516	555	594	633	672	711	750	789	828	867	906	945	984	1023	1062	1101	1140	1179	1218
16'	504	544	584	624	664	704	744	784	824	864	904	944	984	1024	1064	1104	1144	1184	1224	1264
17'	531	572	613	654	695	736	777	818	859	900	941	982	1023	1064	1105	1146	1187	1228	1269	1310
18'	558	600	642	684	726	768	810	852	894	936	978	1020	1062	1104	1146	1188	1230	1272	1314	1356
19'	585	628	671	714	757	800	843	886	929	972	1015	1058	1101	1144	1187	1230	1273	1316	1359	1402
20'	612	656	700	744	788	832	876	920	964	1008	1052	1096	1140	1184	1228	1272	1316	1360	1404	1448
21'	639	684	729	774	819	864	909	954	999	1044	1089	1134	1179	1224	1269	1314	1359	1404	1449	1494
22'	666	712	758	804	850	896	942	988	1034	1080	1126	1172	1218	1264	1310	1356	1402	1448	1494	1540
23'	693	740	787	834	881	928	975	1022	1069	1116	1163	1210	1257	1304	1351	1398	1445	1492	1539	1586
24'	720	768	816	864	912	960	1008	1056	1104	1152	1200	1248	1296	1344	1392	1440	1488	1536	1584	1632

SYMBOLS—ELECTRICAL

	Receptacle outlet, ungrounded
G	Receptacle outlet, grounded
	Duplex receptacle, ungrounded
G	Duplex receptacle, grounded
	Fourplex receptacle, ungrounded
G	Fourplex receptacle, grounded
	Range outlet
	Duplex receptacle, split wired
S	Duplex receptacle and switch
C	Clock receptacle
	Special purpose outlet (designate with subscript)
	Floor special purpose outlet
	Floor receptacle
	Floor duplex receptacle
	Telephone outlet
	Floor telephone outlet
	Fixture outlet, surface incandescent
R	Fixture outlet, recessed incandescent
B	Blanked outlet
F	Fan outlet
J	Junction box

L	Lampholder outlet
LV	Low-voltage controlled outlet
L_{PS}	Lampholder with pull switch
	Fixture outlet, single surface fluorescent
R	Fixture outlet, single recessed fluorescent
	Fixture outlet, multiple surface fluorescent
R	Fixture outlet, multiple recessed fluorescent
MO	Motor outlet
WH	Watt-hour meter
G	Generator
T	Transformer
S	Single-pole switch
S_2	Two-pole switch
S_3	Three-way switch
S_4	Four-way switch
S_D	Door switch
S_K	Key switch
S_P	Switch and pilot lamp
S_{LV}	Low-voltage control switch
S_{LVM}	Low-voltage control master switch
S_T	Time switch

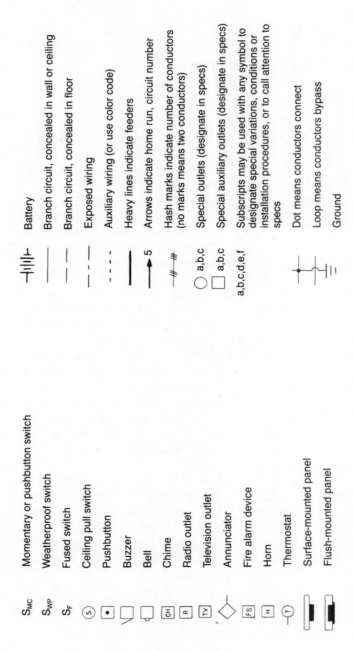

S_{MC}	Momentary or pushbutton switch
S_{WP}	Weatherproof switch
S_F	Fused switch
ⓢ	Ceiling pull switch
•	Pushbutton
	Buzzer
	Bell
CH	Chime
R	Radio outlet
TV	Television outlet
◇	Annunciator
FS	Fire alarm device
H	Horn
Ⓣ	Thermostat
	Surface-mounted panel
	Flush-mounted panel
⊣⊢⊣⊢	Battery
	Branch circuit, concealed in wall or ceiling
	Branch circuit, concealed in floor
	Exposed wiring
	Auxiliary wiring (or use color code)
	Heavy lines indicate feeders
↑5	Arrows indicate home run, circuit number
⧸⧸	Hash marks indicate number of conductors (no marks means two conductors)
◯ a,b,c	Special outlets (designate in specs)
☐ a,b,c	Special auxiliary outlets (designate in specs)
a,b,c,d,e,f	Subscripts may be used with any symbol to designate special variations, conditions or installation procedures, or to call attention to specs
	Dot means conductors connect
	Loop means conductors bypass
	Ground

Glossary

air-dried—lumber that has been piled in yards or sheds for any length of time. For the United States as a whole the minimum moisture content of thoroughly air-dried lumber is 12 to 15%; the average is somewhat higher.

airway—a space between roof insulation and roof boards for movement of air.

alligatoring—coarse checking pattern characterized by a slipping of the new coating over the old coating to the extent that the old coating can be seen through the fissures.

anchor bolts—bolts to secure a wooden sill to concrete or masonry floor or wall.

apron—flat member of the inside trim of a window placed against the wall immediately beneath the stool.

areaway—open subsurface space adjacent to a building used to admit light or air or as a means of access to a basement or cellar.

asphalt—most native asphalt is a residue from evaporated petroleum. It is insoluble in water but is soluble in gasoline and melts when heated. Used widely in building for waterproofing roof coverings of many types, exterior wall coverings, flooring tile, and the like.

astragal—a molding, attached to one of a pair of swinging doors, against which the other door strikes.

attic ventilators—in home building, openings in gables or ventilators in the roof; also, mechanical devices to force ventilation by the use of power-driven fans.

backband—molding used on the side of a door or window casing for ornamentation or to increase the width of the trim.

backfill—the replacement of excavated earth into a pit, trench, or against a structure.

balusters—small spindles or members forming the main part of a railing for stairway or balcony, fastened between a bottom and top rail.

base or baseboard—board placed against the wall around a room next to the floor to finish properly between floor and plaster.

base cap—molding used to trim the upper edge of interior baseboard.

base shoe—molding used next to the floor on interior baseboard. Sometimes called a carpet strip.

batten—narrow strips of wood or metal used to cover joints.

batter board—one of a pair of horizontal boards nailed to posts set at the corners of an excavation, used to indicate the desired level, also as a fastening for stretched strings to indicate outlines of foundation walls.

bay window—any window space projecting outward from the walls of a building, either square or polygonal in plan.

beam—a structural member transversely supporting a load.

bearing partition—a partition that supports any vertical load in addition to its own weight.

bearing wall—a wall that supports any vertical load in addition to its own weight.

bed molding—a molding in an angle, as between the overhanging cornice, or eaves, of a building and the side walls.

bevel—one side of a piece is on a bevel with another when the angle between the two sides is greater or less than 90 degrees.

blinds—light wood sections in the form of doors to close over windows to shut out light, give protection, or add temporary insulation. Commonly used now for ornamental purposes, in which case they are fastened rigidly to the building.

blind-nailing—nailing in such a way that the nailheads are not visible on the face of the work.

blind stop—a rectangular molding, usually 3/4 by 1 3/8 inches or more, used in the assembly of a window frame.

blue stain—a bluish or grayish discoloration of the sapwood caused by the growth of certain moldlike fungi on the surface and in the interior of the piece, made possible by the same conditions that favor the growth of other fungi.

bodied linseed oil—linseed oil in which enough lead, manganese, or cobalt salts have been incorporated to make the oil harden more rapidly when spread in thin coatings.

boiled linseed oil—linseed oil that has been thickened in viscosity by suitable processing with heat or chemicals. Bodied oils are obtainable in a great range in viscosity from a little greater than that of raw oil to just short of a jellied condition.

bolster—a short horizontal timber resting on the top of a column for the support of beams or girders.

boston ridge—a method of applying asphalt or wood shingles as a finish at the ridge or hips of a roof.

brace—an inclined piece of framing lumber used to complete a triangle, and thereby to stiffen a structure.

braced framing—the system of framing buildings by which all vertical structural elements of the bearing walls and partitions extend for one story only. Corner posts excepted.

brick veneer—a facing of brick laid against frame or tile wall construction.

bridging—small wood or metal members that are inserted in a diagonal position between the floor joists to act both as tension and compression members for the purpose of bracing the joists and spreading the action of loads.

buck—often used in reference to rough frame opening members. Door bucks used in reference to metal door frame.

building code—the legal requirements set up by various governing agencies covering the minimum requirements for all types of construction.

built-up roof—a roofing composed of three to five layers of rag felt or jute saturated with coal tar, pitch, or asphalt. The top is finished with crushed slag or gravel. Generally used on flat or low-pitched roofs.

butt joint—the junction where the ends of two timbers or other members meet in a square-cut joint.

cabinet—a shop- or job-built unit for kitchens. Cabinets often include combinations of drawers, doors, etc.

cant strip—a wedge or triangular-shaped piece of lumber used at gable ends under shingles or at the junction of the house and a flat deck under the roofing.

cap—the upper member of a column, pilaster, door cornice, molding, etc.

casing—wide moldings of various widths and thicknesses used to trim door and window openings.

casement frames/sash—frames of wood or metal enclosing part or all of the sash, which may be opened by means of hinges affixed to the vertical edges.

casing—wide moldings of various widths and thicknesses used to trim door and window openings.

cement, Keene's—the whitest finish plaster obtainable that produces a wall of extreme durability. Because of its density it excels for a wainscoting plaster for bathrooms and kitchens and is also used for the finish coat in auditoriums, public buildings, and other places where walls will be subjected to unusually hard wear or abuse.

checking—fissures that with age appear in many exterior paint coatings, at first superficial, but which in time may penetrate entirely through the coating.

checkrails—meeting rails that are thicker than a window to fill the opening between the top and bottom sash made by the parting stop in the frame. They are usually beveled.

collar beam—a beam connecting pairs of opposite roof rafters above the attic floor.

column—in architecture: a perpendicular supporting member, circular or rectangular in section, usually consisting of a base, shaft, and capital. In engineering: a structural compression member, usually vertical, supporting loads acting on or near and in the direction of its longitudinal axis.

combination doors—combination doors have an inside removable section so that the same frame serves for both summer and winter protective devices. A screen is inserted in warm weather to make a screen door, and a glazed or a glazed-and-wood-paneled section in winter to make a storm door. The inconvenience of handling a different door in each season is eliminated.

concrete, plain—concrete without reinforcement, or reinforced only for shrinkage or for temperature changes.

condensation—in a building: beads or drops of water, and frequently frost in extremely cold weather, that accumulate on the inside of the exterior covering of a building when warm, moisture-laden air from inside reaches a point where the temperature no longer permits the air to sustain the moisture it holds. Use of louvers or attic ventilators will reduce moisture condensation in attics.

conduit, electrical—a pipe, usually metal, in which wire is installed.

construction, drywall—a type of construction in which the interior wall finish is applied in a dry condition, generally in the form of sheet materials rather than plaster.

construction, frame—a type of construction in which the structural parts are of wood or dependent upon a wood frame for support. In codes, if brick or other incombustible material is applied to the exterior walls, the classification of this type of construction is usually unchanged.

coped joint—see scribing.

corbel out—to build out one or more courses of brick or stone from the face of a wall, to form a support for timbers.

corner bead—a strip of formed galvanized iron, sometimes combined with a strip of metal lath, placed on corners before plastering to reinforce them. Also, a strip of wood finish three-quarters-round or angular, placed over a plastered corner for protection.

corner boards—used as trim for the external corners of a house; frame structure against which the ends of the siding are finished.

corner braces—diagonal braces let into studs to reinforce corners of frame structures.

cornerite—metal-mesh lath cut into strips and bent to a right angle. Used in interior corners of walls and ceilings on lath to prevent cracks in plastering.

cornice—decorative element made up of molded members usually placed at or near the top of an exterior or interior wall.

cornice return—that portion of the cornice that returns on the gable end of a house.

counterflashing—a flashing usually used on chimneys at the roofline to cover shingle flashing and to prevent moisture entry.

cove molding—a three-sided molding with concave face, used whenever small angles are to be covered.

crawl space—a shallow space below the living quarters of a house. It is generally not excavated or paved and is often enclosed for appearance by a skirting or facing material.

cricket—a small drainage-diverting roof structure of single or double slope, placed at the junction of larger surfaces that meet at an angle.

crown molding—a molding used on cornice or wherever a large angle is to be covered.

dado—a rectangular groove in a board or plank. In interior decoration, a special type of wall treatment.

decay—disintegration of wood or other substance as a result of fungi.

decibel—equivalent to the smallest change in sound energy that the average ear can detect.

deck paint—an enamel with a high degree of resistance to mechanical wear, designed for use on such surfaces as porch floors.

density—the mass of substance in a unit volume. When expressed in the metric system, it is numerically equal to the specific gravity of the same substance.

dimension—see lumber, dimension.

direct nailing—to nail perpendicular to the initial surface or to the junction of the pieces joined. Also termed face nailing.

doorjamb, interior—the surrounding case into which and out of which a door closes and opens. It consists of two upright pieces, called jambs, and a head, fitted together and rabbeted.

dormer—an internal recess, the frame of which projects from a sloping roof.

downspout—pipe, usually of metal, for carrying rainwater from roof gutters.

dressed and matched (tongue and groove)—boards or planks machined in such a manner that there is a groove on one edge and a corresponding tongue on the other.

drier, paint—mostly oil-soluble soaps of such metals as lead, manganese, or cobalt, which, in small proportions, hasten the oxidation and hardening (drying) of the drying oils in paints.

drip—(a) member of a cornice or horizontal exterior-finish course that has a projection beyond the other parts for throwing off water. (b) a groove in the under side of a sill to cause water to drop off on the outer edge, instead of drawing back and running down the face of the building.

drip cap—a molding placed on the exterior top side of a door or window that causes water to drip beyond the outside of the frame.

ducts—in a house, usually round or rectangular metal pipes for distributing warm air from the heating plant to rooms, or air from a conditioning device. Ducts are also made of composition materials.

eaves—the margin or lower part of a roof projecting over the wall.

expanded shale—lightweight aggregate made by expanding shale or clay to form a strong, lightweight aggregate for concrete and concrete masonry units.

expansion joint—a bituminous fiber strip used to separate blocks or units of concrete to prevent cracking due to expansion as a result of temperature changes.

facia or fascia—a flat board, band, or face, used by itself, but more often in combination with moldings. Often located at the outer face of the cornice.

filler (wood)—a heavily pigmented preparation used for filling and leveling off of the pores in open-pored woods.

fire-resistive—in the absence of a specific ruling, applies to materials for construction not combustible in the temperatures of ordinary fires and that will withstand such fires without serious impairment of their usefulness for at least 1 hour.

fire-retardant chemical—a chemical or preparation of chemicals used to reduce flammability or to retard spread of flame.

fire stop—a solid, tight closure of a concealed space, placed to prevent the spread of fire and smoke through such a space.

flagstone (flagging or flags)—flat stones, from 1 to 4 inches thick, used for rustic walks, steps, floors, etc. Usually sold by the ton.

flashing—sheet metal or other material used in roof and wall construction to protect a building from water seepage.

flat paint—an interior paint that contains a high proportion of pigment, and dries to a flat or lusterless finish.

flitch plate—a thin strip of steel or other strong material inserted between two planks to strengthen them.

flue—the space or passage in a chimney through which smoke, gas, or fumes ascend. Each passage is called a flue, which, together with any others and the surrounding masonry, make up the chimney.

flue lining—fire clay or terra-cotta pipe, round or square, usually made in all of the ordinary flue sizes and in 2-foot lengths. Used for the inner lining of chimneys with the brick or masonry work around the outside. Flue lining should run from the concrete footing to the top of the chimney cap. Figure a foot of flue lining for each foot of chimney.

footing—the spreading course or courses at the base or bottom of a foundation wall, pier, or column.

foundation—the supporting portion of a structure below the first-floor construction, or below grade, including the footings.

framing, balloon—a system of framing a building in which all vertical structural elements of the bearing walls and partitions consist of single pieces extending from the top of the soleplate to the roofplate and to which all floor joists are fastened.

framing, platform—a system of framing a building in which floor joists of each story rest on the top plates of the story below or on the foundation sill for the first story, and the bearing walls and partitions rest on the subfloor of each story.

frieze—any sculptured or ornamental band in a building. Also the horizontal member of a cornice set vertically against the wall.

frostline—the depth of frost penetration in soil. This depth varies in different parts of the country. Footings should be placed below this depth to prevent movement.

fungi, wood—microscopic plants that live in damp wood and cause mold, stain, and decay.

fungicide—a chemical that is poisonous to fungi.

furring—strips of wood or metal applied to a wall or other surface to even it, to form an air space, or to give the wall an appearance of greater thickness.

gable—that portion of a wall contained between the slopes of a double-sloped roof, or that portion contained between the slope of a single-sloped roof and a line projected horizontally through the lowest elevation of the roof construction.

gable end—an end wall having a gable.

girder—a large or principal beam used to support concentrated loads at isolated points along its length.

gloss enamel—a finishing material made of varnish and sufficient pigments to provide opacity and color, but little or no pigment of low opacity. Such an enamel forms a hard coating with maximum smoothness of surface and a high degree of gloss.

gloss (paint or enamel)—a paint or enamel that contains a relatively low proportion of pigment and dries to a sheen or luster.

grain—the direction, size, arrangement, appearance, or quality of the fibers in wood.

grain, edge (vertical)—edge-grain lumber has been sawed parallel to the pith of the log and approximately at right angles to the growth rings; i.e., the rings form an angle of 45 degrees or more with the surface of the piece.

grain, flat—flat-grain lumber has been sawed parallel to the pith of log and approximately tangent to the growth rings, i.e., the rings form an angle of less than 45 degrees with the surface of the piece.

grain, quartersawn—another term for edge grain.

grounds—strips of wood, of same thickness as lath and plaster, that are attached to walls before the plastering is done. Used around windows, doors, and other openings as a plaster stop, and in other places for attaching baseboards or other trim.

grout—mortar made of such consistency by the addition of water that it will flow into the joints and cavities of the masonry work and fill them solid.

gutter or eave trough—a shallow channel or conduit of metal or wood set below and along the eaves of a house to catch and carry off rainwater from the roof.

gypsum lath—paper-covered gypsum board used for a base for plasters.

gypsum plaster—gypsum formulated to be used with the addition of sand and water for base-coat plaster.

gypsum sheathing—gypsum board usually, tongue and grooved, covered with a weatherproofed paper for use as exterior sheathing.

header—(a) a beam placed perpendicular to joists and to which joists are nailed in framing for chimney, stairway, or other opening. (b) a wood lintel.

hearth—the floor of a fireplace, usually made of brick, tile, or stone.

heartwood—the wood extending from the pith to the sapwood, the cells of which no longer participate in the life processes of the tree.

hip—the external angle formed by the meeting of two sloping sides of a roof.

hip roof—roof that rises by inclined planes from all four sides of a building.

humidifier—a device designed to discharge water vapor into a confined space for the purpose of increasing or maintaining the relative humidity in an enclosure.

I beam—a steel beam with a cross section resembling the letter "I."

insulating board or fiberboard—a low-density board made of wood, sugarcane, cornstalks, or similar materials, formed by a felting process, dried and usually pressed to thicknesses $1/2$ and $25/32$ inch.

insulation, building—any material high in resistance to heat transmission that, when placed in the walls, ceilings, or floors of a structure, will reduce the rate of heat flow.

jamb—the side post or lining of a doorway, window, or other opening.

joint—the space between the adjacent surfaces of two members or components joined and held together by nails, glue, cement, mortar, or other means.

joint compound—a powder that is usually mixed with water and used for joint treatment in gypsum-wallboard finish. Available ready-mixed.

joist—one of a series of parallel beams used to support floor and ceiling loads, and supported in turn by larger beams, girders, or bearing walls.

k-factor—k-factor is a measure of the heat or cold conductive value of a material. It is based on the B.T.U. per hour per square foot per degree F per inch of thickness.

knot—that portion of a branch or limb that has become incorporated in the body of a tree.

landing—a platform between flights of stairs or at the termination of a flight of stairs.

lap siding—see siding, beveled.

lath—a building material of wood, metal, gypsum, or insulating board that is fastened to the frame of a building to act as a plaster base.

lattice—an assemblage of wood or metal strips, rods, or bars made by crossing them to form a network.

leader—see downspout.

ledger strip—a strip of lumber nailed along the bottom of the side of a girder on which joists rest.

light—space in a window sash for a single pane of glass. Also, a pane of glass.

lintel—a horizontal structural member that supports the load over an opening such as a door or window.

lookout—a short wood bracket or cantilever to support an overhanging portion of a roof, usually concealed from view.

louver—an opening with a series of horizontal slats arranged to permit ventilation but to exclude rain, sunlight, or vision. See also attic ventiltors.

lumber—lumber is the product of the sawmill and planing mill not further manufactured other than by sawing, resawing, and passing lengthwise through a standard planing machine, crosscut to length and matched.

lumber, boards—yard lumber less than 2 inches thick and 2 or more inches wide.

lumber, dimension—yard lumber from 2 inches to but not including 5 inches thick, and 2 or more inches wide. Includes joists, rafters, studding, plank, and small timbers.

lumber, dressed size—the dimensions of lumber after shrinking from the green dimension and after planing, less than the nominal or rough size. For example, a 2 by 4 stud actually measures $1^1/_2 \times 3^1/_2$ inches.

lumber, matched—lumber that is edge-dressed and shaped to make a close tongue-and-groove joint at the edges or ends when laid edge to edge or end to end.

lumber, shiplap—lumber that is edge-dressed to make a close rabbeted or lapped joint.

lumber, timbers—yard lumber 5 or more inches in least dimension. Includes beams, stringers, posts, caps, sills, girders, and purlins.

lumber, yard—lumber of those grades, sizes, and patterns which are generally intended for ordinary construction, such as framework and rough coverage of houses.

mantel—the shelf above a fireplace. Originally referred to the beam or lintel supporting the arch above the fireplace opening. Used also in referring to the entire finish around a fireplace, covering the chimney breast in front and sometimes on the sides.

masonry—stone, brick, concrete, hollow-tile, concrete-block, gypsum-block, or other similar building units or materials or a combination of the same, bonded together with mortar to form a wall, pier, buttress, or similar mass.

masonry fill—insulating material

used to fill the cores or open spaces in masonry construction for insulation purposes.

masonry reinforcing—lateral steel reinforcing rods or mesh used between courses in masonry construction.

metal lath—sheets of metal that are slit and drawn out to form openings on which plaster is spread.

millwork—most building materials made of finished wood and manufactured in millwork plants and planing mills are included under the term ''millwork.''It includes such items as inside and outside doors, window and doorframes, blinds, porchwork, mantels, panelwork, stairways, moldings, and interior trim. It does not include flooring, ceiling, or siding.

miter—the joining of two pieces at an angle that bisects the angle of junction.

moisture content of wood—weight of the water contained in the wood, usually expressed as a percentage of the weight of the oven-dry wood.

molding—usually patterned strips, used to provide ornamental variation of outline or contour, whether projections or cavities, such as cornices, bases, window and doorjambs, and heads.

mortise—a slot cut into a board, plank, or timber, usually edgewise, to receive tenon of another board, plank, or timber to form a joint.

mullion—a slender vertical bar or pier forming a division between panels or window groups, screens, or similar frames.

muntin—the members dividing the glass or openings of sash, doors, etc.

natural finish—a transparent finish, usually a drying oil, sealer, or varnish, applied on wood for protection against soiling or weathering. Such a finish may not seriously alter the original color of the wood or obscure its grain pattern.

newel—a post to which the end of a stair railing or balustrade is fastened. Also, any post to which a railing or balustrade is fastened.

nonbearing wall—a wall supporting no load other than its own weight.

nosing—the projecting edge of a molding or drip. Usually applied to the projecting molding on the edge of a stair tread.

O.C. (on center)—the measurement of spacing for studs, rafters, joists, etc. in a building from center of one member to the center of the next member.

O.G. (or ogee)—a molding with a profile in the form of a letter S; having the outline of a reversed curve.

plywood—a piece of wood made of three or more layers of veneer joined with glue and usually laid with the grain of adjoining plies at right angles. Almost always an odd number of plies are used to provide balanced construction.

polyethylene—a thin plastic film in transparent or black. Used as a vapor barrier in building.

panel—a large, thin board or sheet of lumber, plywood, or other material. A thin board with all its edges inserted in a groove of a surrounding frame of thick material. A portion of a flat surface recessed or sunk below the surrounding area, distinctly set off by molding or some other decorative device. Also, a section of floor, wall, ceiling or roof, usually prefabricated and of large size, handled as a single unit in assembly and erection.

paper, building—a general term for papers, felts, and similar sheet materials used in buildings without reference to their properties or uses.

paper, sheathing—a building material, generally paper or felt, used in wall and roof construction as a protection against the passage of air and sometimes moisture.

parting stop or strip—a small wood piece used in the side and head jambs of double-hung windows to separate upper and lower sash.

partition—a wall that subdivides spaces within any story of a building.

penny—as applied to nails, originally indicated the price per hundred. The term now serves as a measure of nail length and is abbreviated by the letter "d."

perlite—a lightweight, artificially expanded white glasslike material or aggregate used for lightweight concrete, plaster aggregate, acoustical correction, and insulation.

perm—a unit of permeance; the passage of one grain of water vapor per square foot per hour per inch of mercury.

permeability—the property of a material that allows the passage of water vapor. It is equal to the permeance of a 1-inch thickness of the material, and is measured in perm-inches.

permeance—the ratio of the transmission of water vapor of a material to the difference in vapor pressure between the two opposite surfaces of the material.

pier—a column of masonry, usually rectangular in horizontal cross section, used to support other structural members.

pigment—a powdered solid in suitable degree of subdivision, for use in paint or enamel.

pitch—the incline or rise of a roof. Pitch is expressed in inches or rise per foot of run, or by the ratio of the rise to the span.

pitch pocket—an opening extending parallel to the annual rings of growth, that usually contains, or has contained, either solid or liquid pitch.

pith—the small, soft core at the original center of a tree around which wood formation takes place.

plastic foam—an expanded plastic material used for insulation purposes.

plate—(1) a horizontal structural member placed on a wall or supported on posts, studs, or corbels to carry the trusses of a roof or to carry the rafters directly. (2) A shoe, or base member, as of a partition or other frame. (3) A small, flat member placed on or in a wall to support girders, rafters, etc.

plinth—a slab block or projection under a column or in a wall to form a projecting course.

plough—to cut a groove, as in a plank.

plumb—exactly perpendicular; vertical.

ply—a term to denote the number of thicknesses or layers of roofing felt, veneer in plywood, or layers in built-up materials.

plywood—a piece of wood made of three or more layers of veneer joined with glue and usually laid with the grain of adjoining plies at right angles. Almost always an odd number of plies are used to provide balanced construction.

polyethylene—a thin plastic film in transparent or black. Used as a vapor barrier in building.

porch—a floor extending beyond the exterior walls of a building. It may be covered and enclosed or unenclosed.

pores—wood cells of comparatively large diameter that have open ends and are set one above the other to form continuous tubes. The openings of the vessels on the surface of a piece of wood are referred to as pores.

preacher—the term used to designate a U-shaped piece of wood used by masons to hold their line in place on a wall.

preservative—any substance that, for a reasonable length of time, will prevent the action of wood-destroying fungi, borers of various kinds, and similar destructive life when the wood has been properly coated or impregnated with it.

prestressed—a process of preparing concrete slabs and beams for extra strengths by pouring the mix over tightly-drawn special steel wire rope or rods which are later released to provide strong dense concrete.

primer—the first coat of paint in a paint job that consists of two or more coats; also the paint used for such a first coat.

pumice—naturally expanded perlite rock used as a lightweight aggregate for concrete and concrete masonry.

put-log—a cross piece in scaffolding.

putty—a type of cement usually made of whiting and boiled linseed oil, beaten or kneaded to the consistency of dough and used in sealing glass in sash, filling small holes and crevices in wood, and for similar purposes.

quarter round—molding that presents a profile of a quarter circle.

R value—the R value, or resistance value, is a comparative or additive measure of the ability of a building material or composite section to resist heat change or heat flow through it. The higher the value, the greater the resistance, thermal efficiency, and insulating value.

rabbet—a rectangular longitudinal groove cut in the corner of a board or other piece of material.

radiant heating—a method of heating, usually consisting of coils or pipes placed in the floor, wall, or ceiling.

rafter—one of a series of structural members of a roof designed to support roof loads. The rafters of a flat roof are sometimes called roof joists.

rafter, hip—a rafter that forms the intersection of an external roof angle.

rafter, jack—a rafter that spans the distance from a wallplate to a hip or from a valley to a ridge.

rafter, valley—a rafter that forms the intersection of an internal roof angle.

rail—a horizontal bar or timber of wood or metal extending from one post or support to another as a guard or barrier in a fence, balustrade, staircase, etc. Also, the cross or horizontal members of the framework of a sash, door, blind, or any paneled assembly.

rake—the trim members that run parallel to the roof slope and from the finish between wall and roof.

raw linseed oil—the crude product obtained from flaxseed.

reflective insulation—sheet material with one or both surfaces of comparatively low heat emissivity that, when used in building construction so that the surfaces face air spaces, reduces the radiation across the air space.

reinforcing—steel rods or metal fabric to place in concrete slabs, beams, or columns to increase their strength.

resin-emulsion paint—paint, the vehicle (liquid part) which consists of resin or varnish dispersed in fine droplets in water. (Analo-gous to cream, which consists of butterfat dispersed in water.

relative humidity—the amount of water vapor expressed as a percentage of the maximum quantity that could be present in the atmosphere at a given temperature. (The actual amount of water vapor that can be held in space increases with the temperature.)

ribbon—a narrow board let into the studding to add support to joists.

ridge—the horizontal line at the junction of the top edges of two sloping roof surfaces. The rafters at both slopes are nailed at the ridge.

ridge board—the board placed at the ridge of the roof to support the upper ends of the rafters.

rise—the height of a roof rising in horizontal distance (run) from the outside face of a wall supporting the rafters or trusses to the ridge of the roof. In stairs, the perpendicular height of a step or flight of steps.

riser—each of the vertical boards closing the spaces between the treads of stairways.

roll roofing—roofing material, composed of fiber and saturated with asphalt, that is supplied in rolls containing 108 square feet in 36-inch widths. It is generally furnished in weights of 55 to 90 pounds per roll.

roof sheathing—boards or sheet material fastened to the roof rafters on which the shingle or other roof covering is laid.

rubber-emulsion paint—paint, the vehicle of which consists of rubber or synthetic rubber dispersed in fine droplets in water.

run—in reference to roofs, the horizontal distance from the face of a wall to the ridge of the roof. Referring to stairways, the net width of a step; also the horizontal distance covered by a flight of steps.

sapwood—the outer zone of wood, next to the bark. In the living tree it contains some living cells (the heartwood contains none), as well as dead and dying cells. In most species, it is lighter colored than the heartwood. In all species, it is lacking in decay resistance.

sash—single frame containing one or more lights of glass.

sash balance—a device, usually operated with a spring, designed to counterbalance window sash. Use of sash balances eliminates the need for sash weights, pulleys, and sash cord.

saturated felt—a felt that is impregnated with tar or asphalt.

scratch coat—the first coat of plaster, which is scratched to form a bond for the second coat.

scribing—fitting woodwork to an irregular surface.

sealer—a finishing material, either clear or pigmented, that is usually applied directly over uncoated wood for the purpose of sealing the surface.

seasoning—removing moisture from green wood in order to improve its serviceability.

semigloss paint or enamel—a paint or enamel made with a slight insufficiency of nonvolatile vehicle so that its coating, when dry, has some luster but is not very glossy.

shake—a handsplit shingle, usually edge grained.

sheathing—the structural covering, usually wood boards, plywood, or wallboards, placed over exterior studding or rafters of a structure.

sheathing paper—see paper, sheathing.

shellac—transparent coating made by dissolving lac, a resinous secretion of the lac bug (a scale insect that thrives in tropical countries, especially India), in alcohol.

shingles—roof covering of asphalt, wood, tile, slate, or other material cut to stock lengths, widths, and thicknesses.

shingles, siding—various kinds of shingles, some especially designed, that can be used as the exterior sidewall covering for a structure.

shiplap—see lumber, shiplap.

shutters—ventilated or louvered panels usually hinged to cover windows. Also used as decorative units fastened to the outside walls adjacent to windows.

siding—the finish covering of the outside wall of a frame building, whether made of weatherboards, vertical boards with battens, shingles, or other material.

siding, bevel (lap siding)—used as the finish siding on the exterior of a house or other structure. It is usually manufactured by resawing dry square-surfaced boards diagonally to produce two wedge-shaped pieces. These pieces commonly run from $3/16$ inch thick on the thin edge to $1/2$ to $3/4$ inch thick on the other edge, depending on the width of the siding.

siding, drop—usually $3/4$ inch thick and 6 inches wide, machined into various patterns. Drop siding has tongue-and-groove joints, is heavier, and has more structural strength than other types. It is frequently used on buildings that require no sheathing, such as garages and barns.

sill—the lowest member of the frame of a structure, resting on the foundation and supporting the uprights of the frame. The member forming the lower side of an opening, as a door sill, window sill, etc.

soffit—the underside of the members of a building, such as staircases, cornices, beams, and arches, relatively minor in area as compared with ceilings.

soil cover (ground cover)—a light roll roofing used on the ground of crawlspaces to minimize moisture permeation of the area.

soil stack—a general term for the vertical main of a system of soil, waste, or vent piping.

sole or soleplate—a member, usually a 2 by 4, on which wall and partition studs rest.

span—the distance between structural supports such as walls, columns, piers, beams, girders, and trusses.

specific gravity—the specific gravity of a substance is its weight as compared with the weight of an equal bulk of pure water. For making specific gravity determina- tions the temperature of the water is usually taken at 62° F., when 1 cubic foot of water weighs 62.355 pounds. Water is at its greatest density at 39.2° F. or 4° Centi- grade.

splash block—a small masonry block laid with the top close to the ground surface to receive roof drainage and to carry it away from the building.

square—a unit of measure—100 square feet—usually applied to roofing material. Sidewall cover- ings are often packed to cover 100 square feet and are sold on that basis.

stain, shingle—a form of oil paint, very thin in consistency, intended for coloring wood with rough sur- faces, like shingles, without form- ing a coating with a significant thickness or gloss.

stair carriage—a stringer for steps on stairs.

stair landing—a platform between flights of stairs or at the termina- tion of a flight of stairs.

stair rise—vertical distance from the top of one stair tread to the top of the next one above.

stool—the flat, narrow shelf forming the top member of the interior trim at the bottom of a window.

storm sash or storm window—an extra window usually placed on the outside of an existing window as additional protection against cold weather.

story—part of a building between any floor and the floor or roof next above.

string, stringer—a timber or other support for cross members. In stairs, the support on which the stair treads rest; also stringboard.

stucco—most commonly refers to an outside plaster made with Portland cement as its base.

stud—one of a series of slender wood or metal structural members placed as supporting elements in walls and partitions. (Plural: studs or studding.)

subfloor—boards or sheet material laid on joists over which a finish floor is to be laid.

tail beam—a relatively short beam or joist supported in a wall on one end and by a header on the other.

termites—insects that superficially resemble ants in size, general appearance, and habit of living in colonies; hence, they are frequently called "white ants." Subterranean termites do not establish themselves in buildings by being carried in with lumber but by entering from ground nests after the building has been constructed. If unmolested, they eat out the woodwork, leaving a shell of sound wood to conceal their activities, and damage may proceed so far as to cause collapse of parts of a structure before discovery. There are about 56 species of termites known in the United States, but the two major species, classified from the manner in which they attack wood, are ground-inhabiting or subterranean termites, the most common, and dry wood termites, found almost exclusively along the extreme southern border and the Gulf of Mexico in the United States.

termite shield—a shield, usually of noncorrodible metal, placed in or on a foundation wall or other mass of masonry or around pipes to prevent passage of termites.

therm—100,000 Btu (British thermal unit) of gross heat value.

threshold—a strip of wood or metal beveled on each side and used above the finished floor under outside doors.

ties—metal rods or heavy wires used to hold concrete forms in place during the pouring of walls.

tie wire—wire used for above purpose, also used in tying metal lath to columns and supports. Also, wire mesh for reinforcing concrete, plaster or stucco.

toenailing—to drive a nail at a slant with the initial surface in order to permit it to penetrate into a second member.

tread—the horizontal board in a stairway on which the foot is placed.

trim—the finish materials in a building, such as moldings, applied around openings (window trim, door trim) or at the floor and ceiling of rooms (baseboard, cornice, picture molding).

trimmer—A beam or joist to which a header is nailed in framing for a chimney, stairway, or other opening.

truss—a frame or jointed structure designed to act as a beam of long span, while each member is usually subjected to longitudinal stress only, either tension or compression.

turpentine—a volatile oil used as a thinner in paints and as a solvent in varnishes. Chemically, it is a mixture of terpenes.

U factor—U is the symbol for the rate of thermal transmittance through a building material or composite section. In assessing discrete or composite materials, it is referred to as the heat transmission coefficient.

undercoat—a coating applied prior to the finishing or top coats of a paint job. It may be the first of two or the second of three coats. Also refers to priming coat.

valley—the internal angle formed by the junction of two sloping sides of a roof.

vapor barrier—material used to retard the flow of vapor or moisture into walls and thus prevent condensation within them. There are two types of vapor barriers, the membrane that comes in rolls and is applied as a unit in the wall or ceiling construction, and the paint type, which is applied with a brush. The vapor barrier must be a part of the heated/cooled side of the wall.

varnish—a thickened preparation of drying oil or drying oil and resin suitable for spreading on surfaces to form continuous, transparent coatings, or for mixing with pigments to make enamels.

vehicle—the liquid portion of a finishing material; consisting of the binder (nonvolatile) and volatile thinners.

veneer—thin sheets of wood. A stucco, plaster or brick coating.

vent—a pipe installed to provide a flow of air to or from a drainage system or to provide a circulation of air within such system to protect trap seals from siphonage and back pressure.

vermiculite—a mineral closely related to mica, with the faculty of expanding on heating to form lightweight material with insulation quality. Used as bulk insulation and also as aggregate in insulating and acoustical plaster and in insulating concrete floors.

volatile thinner—a liquid that evaporates readily and is used to thin or reduce the consistency of finishes without altering the relative volumes of pigments and nonvolatile vehicles.

wallboard—woodpulp, gypsum, or other materials made into large, rigid sheets that may be fastened to the frame of a building to provide a surface finish.

wall ties—metal units used to tie together masonry units when two types are used in the same wall or when cavity type walls are built. Both round and flat types are used—usually galvanized, copper clad, or copper.

wale—a strip of wood, usually 2 × 4 or heavier, used to reinforce form panels for concrete work used both vertically and horizontally.

wane—bark, or lack of wood or bark from any cause, on edge or corner of a piece.

wash—the upper surface of a member or material when given a slope to shed water.

water repellent—a liquid designed to penetrate into wood and to impart water repellency to the wood.

water table—a ledge or offset on or above a foundation wall, for the purpose of shedding water.

weatherstrip—narrow strips made of metal, or other material, so designed that when installed at doors or windows they will retard the passage of air, water, moisture, or dust around door or window sash.

wood rays—strips of cells extending radially within a tree and varying in height from a few cells in some species to 4 inches or more in oak. The rays serve primarily to store food and to transport it horizontally in the tree.

Index